Algebra
for High School

Book 1 - Linear Equations and Inequalities

Marcel Sincraian, Ph.D.

ISBN: 978-1-7775022-9-4   Electronic Book

ISBN: 978-1-7775022-8-7   Printed Book

Marcel Sincraian Email: msincraian@yahoo.ca

To
My wife

CONTENT

Chapter 1 Cartesian System — 3

1.A. Understanding the transition from numbers to variables — 4

1.B. Ordered pairs - Introduction — 5

PRACTICE — 6

Chapter 2 Straight Lines — 8

2.A. Representing patterns in linear relations — 9

PRACTICE — 10

2.B. Understanding tables of values of linear relationships — 14

PRACTICE — 17

2.C. Understanding graphs of linear relationships — 19

PRACTICE — 20

2.D. The Distance between points — 23

a. Horizontal distance — 23

PRACTICE — 25

b. Vertical distance — 26

PRACTICE — 27

2.D. Distance between points — 28

c. Non-horizontal and Non-vertical distance — 28

PRACTICE — 29

2.D. The Distance between points — 30

d. Midpoint coordinates — 30

PRACTICE — 32

2.E. The Slope of a line — 33

PRACTICE — 35

Chapter 3 Linear Equations — 36

3.A. Linear equations — 37

3.B. Solving Linear Equations — 38

a. Solve one-step linear equations: — 38

PRACTICE — 39

3.B. Solving Linear Equations — 41

b. Solving two-step linear equations with addition and subtraction — 41

PRACTICE — 42

3.B. Solving Linear Equations — 44

c. Solving two-step linear equations with multiplication and division — 44

PRACTICE — 45

3.B. Solving Linear Equations — 47

d. Solving two- step linear equations using distributive property — 47

PRACTICE — 47

3.C. Equation of a straight line — 49

a. Non-vertical and non-horizontal line — 49

PRACTICE — 51

3.D. Straight-line graph — 53

a. From equation to the graph — 53

b. From graph to the equation — 54

PRACTICE — 55

3.E. Special cases of linear equations: — 56

Vertical and horizontal lines — 56

PRACTICE — 57

3.F. Parallel and perpendicular lines — 59

a. Parallel lines — 59

PRACTICE — 60

3.F. Parallel and perpendicular lines — 61

b. Perpendicular lines — 61

PRACTICE — 62

Examples of applications of linear relationships — 63

Chapter 4 Linear Inequalities — **65**

4.A. Express linear inequalities graphically and algebraically — 66

PRACTICE — 67

4.B. Solving one-step linear inequalities — 69

PRACTICE — 70

4.C. Solving multi-step linear inequalities — 72

PRACTICE — 74

4.D. Linear inequalities with two variables — 76

PRACTICE — 78

STEP BY STEP SOLUTIONS — 80

Chapter 1

Cartesian System

..A. Understanding the transition from numbers to variables

Remember that in smaller grades we had:

$+7 = 12$

It was obvious that 5 plus 7 is 12.

Remember as well, that we had the problems where we were asked: what number plus 7 equals 12?

$? + 7 = 12$

Or what number minus 4 equals 5?

$? - 4 = 5$

In Algebra we substitute the unknown number with a letter, say x

So, the above relations will become:

$+7 = 12$

Or,

$-4 = 5$

X or any other letter is called the variable or the unknown.

> Remember:
> The equal sign shows that what is on the left side of the sign is exactly the same amount or value as what is on the right side of the sign

Any expression and operations between expressions that include one or more variables, is part of algebra.

Equal sign concept

The equal sign is fundamental in mathematics. It tells us that whatever number, letter, expression is on the left side of the sign, is exactly the same as what is on the right side of the equal sign.

It is obvious that 5=5. But what is very important to realize here is that the equal sign tells us that the number 5 to the left of = is the same with number 5 to the right of = sign.

Now, if we have:

X+2=3

Here it means that the final value of x+2 has to be the same as 3. We will come back to this concept when we discuss equations and how to solve them.

1.B. Ordered pairs - Introduction

Remember in lower grades when you learned about the number line.

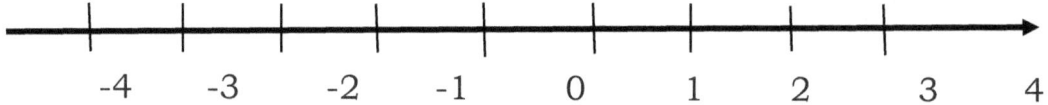

Each number is at a certain distance from the origin 0. Either to the left (negative numbers) or to the right (positive numbers).

In the 17th century Rene Descartes came up with the idea of using two number lines, one horizontal, and the other one vertical. From here, each point in the 2-dimensional plane received a pair of ordered coordinates, x for the horizontal line, and y for the vertical line. (x , y)
This is called the Cartesian System of axes.

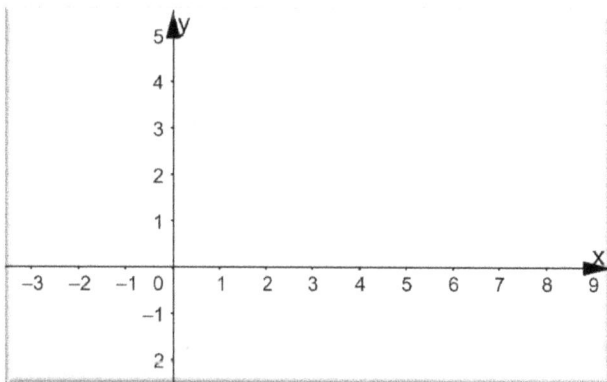

(0,0)

EXAMPLE

If we want to represent any point situated anywhere in the 2-dimensional plane, we will have to give the point the coordinates x and y.
(x, y)
The intersection of the axes is called the system origin or center. The origin has the coordinates zero and zero.

NOTE:
We always have the coordinates in the same order, first the x coordinate, then the y coordinate.
The system has four quadrants.
Quadrant 1 where both coordinates are positive.
Quadrant 2 where x is negative, and y is positive.
Quadrant 3 where both x and y coordinates are negative.
Quadrant 4 where x is positive, and y is negative.

Remember:
The coordinates are written in the same order x first and then y

Q2	Q1
Q3	Q4

Let's represent a few points in the Cartesian system.

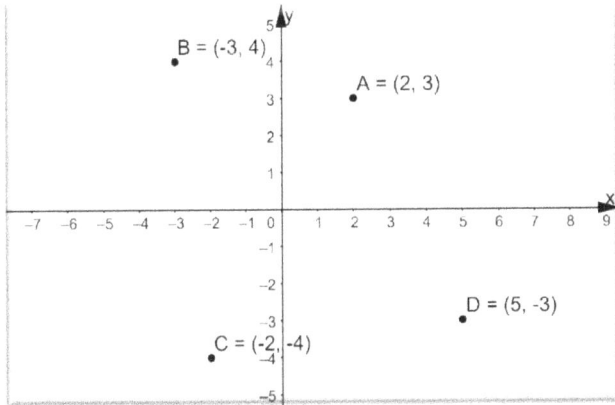

EXAMPLE

Represent the following points.
A(2,3), B(-3,4), C(-2,-4), D(5,-3)
As it can be seen, for point A the x
coordinate equals 2, and the y
coordinate equals 3. For point D, for
example, the x coordinate equals 5,
and the y coordinate equals minus 3.

PRACTICE

Represent the following points on the graph below.

A (1,1)   B (3,5)
C (-2,4)   D (-3,1)
E (-3,-5)   F (-4,0)
G (4,-2)   H (3,-3)

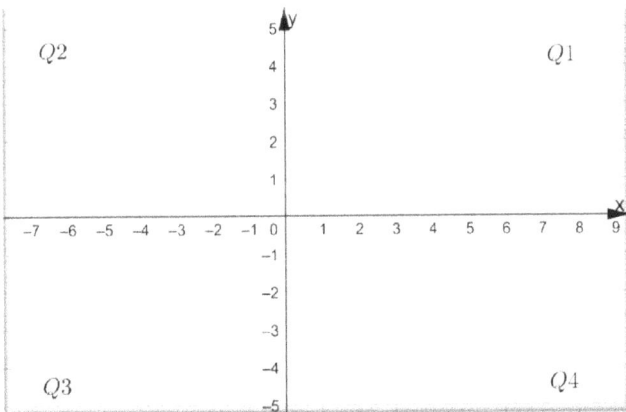

In what quadrants are the points below situated?

A (1,1.   B (3,5)

C (-2,4)   D (-3,1.

E (-3,-5)   F (-4,0)

G (4,-2)   H (3,-3)

Chapter 2

Straight Lines

2.A. Representing patterns in linear relations

Let's suppose we want to count how many sticks we need to build 20 connected triangles.

We can see in the figure below that, in order to build one triangle, we need three sticks.

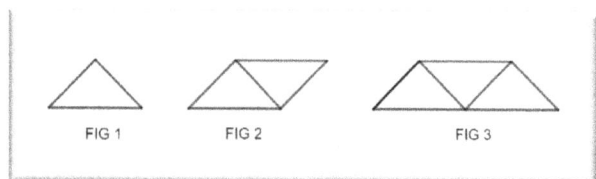

FIG 1   FIG 2   FIG 3

To build two connected triangles, we need 5 sticks. (FIG 2)

To build three connected triangles, we need 7 sticks. (FIG 3)

Let's gather all this information in a table with two columns.

Triangles	Sticks
1	3
2	5
3	7
4	9
20	41

We can see that as the number of triangles increases by 1, the number of sticks increases by 2 with each new connected triangle. If we continue this pattern, we will see that for 20 triangles we will need 41 sticks.

In the figure below, we have the case where we start with one square, then we add

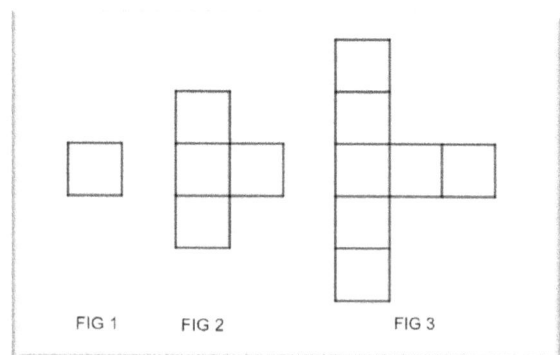

FIG 1   FIG 2   FIG 3

three squares, one above, one to the right and one below.

FIG 3 shows the case when we added another set of three squares to the ones in FIG 2.

How many squares would we need in FIG 7?

Again, we can gather all the information in a table with two columns. The first column contains the figure number. The second column contains the number of squares.

FIG #	Squares
1	1
2	4
3	7
4	10
7	19

We can see that as the FIG's number increases by 1, the number of squares increases by 3 with each new connected triangle. If we continue this pattern, we will see that in FIG 7 we will have 19 squares.

PRACTICE

1) Analyze the pattern shown in the figures below. Find how many houses figure 5 will have.

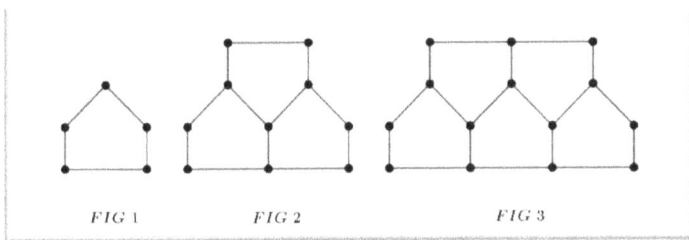

FIG 1   FIG 2   FIG 3

Figure	# Houses
1	1
2	3
3	5
5	

2) Analyze the pattern shown in figures below. Find how many sticks figure 6 will have.

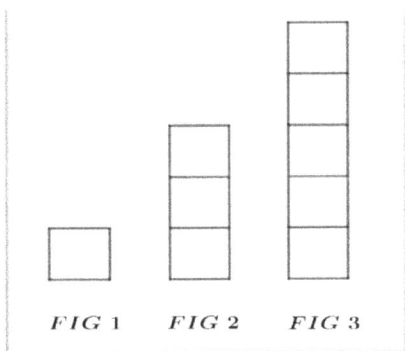

FIG 1   FIG 2   FIG 3

Figure	# Squares
1	1
2	3
3	5
4	7
6	

3) Analyze the pattern shown in figures below. Find how many squares figure 5 will have.

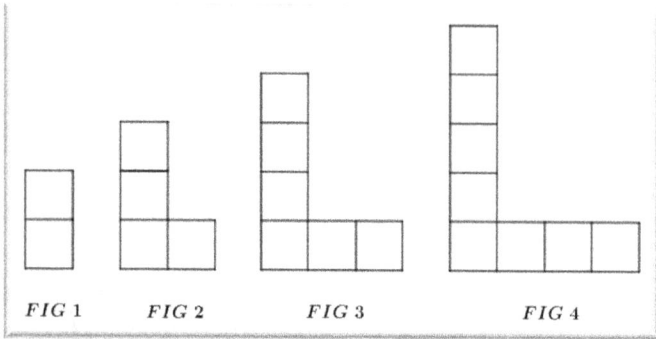

Figure	# Squares
1	2
2	4
3	6
4	8
5	

4) Analyze the pattern shown in figures below. Find how many squares figure 6 will have.

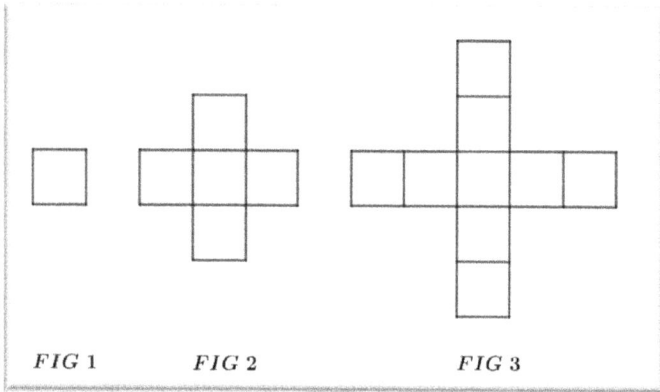

Figure	# Squares
1	1
2	5
3	9
4	13
6	

5) Analyze the pattern shown in figures below. Find how many triangles figure 5 will have.

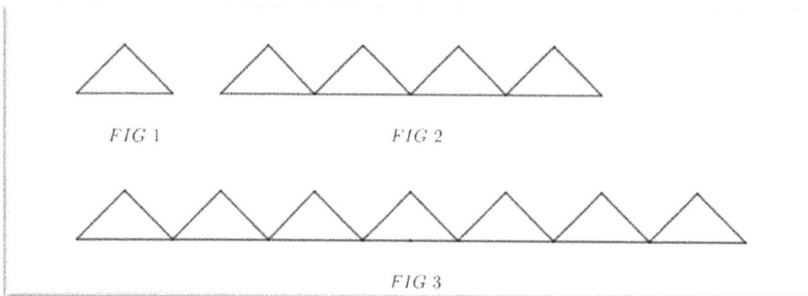

Figure	# Triangles
1	1
2	4
3	7
4	10
5	

6) Analyze the pattern shown in the table below. Find how many figurines the 5th child will have.

Child	# Figurines
1	3
2	7
3	11
4	15
5	

7) Victoria needs to build cardboard cubes. After building the fourth cube, she is left with 31 carboard pieces. Can she finish the tenth cube?

Cube	# Carboard pieces
1	6
2	12
3	18
4	24

8) Analyze the pattern shown in the figures below. Find how many squares the 5th figure will have.

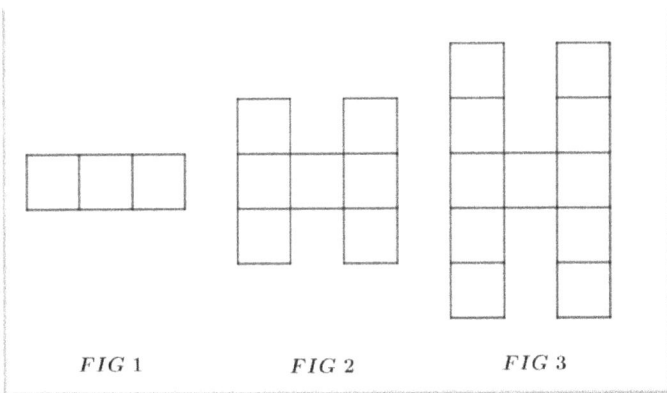

FIG 1   FIG 2   FIG 3

Figure	# Squares
1	3
2	7
3	11
4	15
5	

9) Make a table of values for the first four terms. Start at 4. To get the next term, triple the number and subtract 1.

X	Y=
4	Y=
5	Y=
6	Y=
7	Y=

10) A rental car business has the client pay $40 for the first hour and $5 for every hour after that. How much does it cost to rent the car for 12 hours?

2.B. Understanding tables of values of linear relationships

Let's start with a table of values. The first column represents the x values. The second column represents the y values. Each <u>ordered</u> pair (x, y) represents a point in the system of axes.

Consider the table below.

X	Y
2	3
3	5
4	7
5	9

Remember:
The ordered pair will always be
X FIRST
Y SECOND

What do we notice in each column in the table above?

As successive x values increase by one unit, the successive y values increase with two units all the time.

For example, in the table above, in the x column, first value is 2, then 3, then 4, then 5. The difference between 3 and 2 is one, between 4 and 3 is again one, and so on. In the y column we have 3 as the first value, then 5, then 7, then 9. The difference between 5 and 3 is two. The difference between 7 and 5 is two again, and so on.

The increase in each column is the same with one unit for x and with two units for y respectively. The difference is not always 1 for x and 2 for y. It can be any difference, as long as it is the same for each successive value for x and y respectively.

EXAMPLE

Consider the table below:

X	Y
1	2
2	6
3	10
4	14

In the table above, in the x column, the first value is 1, then 2, then 3, then 4. The difference between 2 and 1 is one, between 3 and 2 is again one, and so on. In the y column we have 2 as the first value, then 6, then 10, then 14. The difference between

6 and 2 is four. The difference between 10 and 6 is four again, the difference between 14 and 10 is four again.

The relation between x and y in the tables

Consider the table below.

X	Y
2	3
3	5
4	7
5	9

As successive x values increase by one unit, the successive y values increase with two units all the time.

For example, in the table above, in the x column, the first value is 2, then 3, then 4, then 5. The difference between 3 and 2 is one, between 4 and 3 is again one, and so on. In the y column we have 3 as the first value, then 5, then 7, then 9. The difference between 5 and 3 is two. The difference between 7 and 5 is two again, and so on. In the column of y we increase by 2 all the time

We can build a relation between x and y.

$y = 4x$

If we consider $x = 1$, then we have $y = 4 * 1 = 4$ which is two units more than 2.

So, to have $y = 4$ we need to subtract 2 to obtain $y = 2$

The general relation here will be $y = 4x - 2$

We can check with different values for x to see if the formula is good.

If $x = 4$ then we have $y = 4 * 4 - 2 = 16 - 2 = 14$, which is exactly the number we were expecting to get. The relation between y and x is indeed $y = 4x - 2$.

So, in order to create the relationship between y and x, we have to follow these steps: supposing that the difference between the two consecutive values in the x column is one:

Step 1

Check the difference between two consecutive values in the y column. (Let's call it A)

Step 2

Form the equation $y = A \times x$

Step 3

Check if for $x = 1$, the value for y at step 2 equals the value we should get in column y

Step 4

We add or subtract any value from $A \times x$ in such a way that we obtain the value of y that corresponds to the value of x in the x column.

Step 5

Check for another value of x to see if we obtain the correct corresponding value of y.

EXAMPLE

Consider the table below:

X	Y
1	7
2	10
3	13
4	16

Step 1

Check the difference between two consecutive values in the y column.

In this case that difference is 3

Step 2

Form the equation $y = 3 \times x$

Step 3

Check if for $x = 1$, the value for y at step 2 equals the value we should get in column y.

For $x = 1$, $y = 3 \times (1) = 3$

Step 4

We add or subtract any value from $A \times x$ in such a way that we obtain the value of y that is beside the value of x in the x column.

The value of y that corresponds to x=1 is 7 not 3.

We have to add 4 units to $y = 3 \times x$ in order to get 7.

The relation between y and x in this case will be:

$y = 3 \times x + 4$

Step 5

Check for another value of x to see if we obtain the correct corresponding value of y.

Let us take $x = 3$

Then we have: $y = 3 \times (3) + 4 = 9 + 4 = 13$ which equals the value of y for $x = 3$.

NOTE

If the first value of x is different than 1, we will take that value whichever it is.

PRACTICE

1) In the relation $B = 4q$, determine B when q is:

a) 2   b) 13

c) -5   d) x+2

2) In the relation $M = 3k - 4$ determine M when k is:

a) 2   b) 17

c) 40   d) y+3

3) Determine the common difference in each linear pattern.

a) 2,5,8,11   b) $\sqrt{3},\ 2\sqrt{3},\ 3\sqrt{3},\ 4\sqrt{3}$

4) Determine the common difference in the pattern below.

$\sqrt{3} + \sqrt{5}, \quad \sqrt{3}, \quad \sqrt{3} - \sqrt{5} \dots \dots \dots$

5) Analyze the table below and write a relation between x and y.

X	1	2	3	4	5
Y	6	9	12	15	18

6) Analyze the table below and write a relation between x and y.

X	1	2	3	4	5
Y	15	10	5	0	-5

7) Analyze the table below and write a relation between x and y.

X	1	2	3	4	5
Y	-5	-1	3	7	11

8) Determine the 20th term of the billow linear pattern

5,9,13,17,21

We create the table:

X (term #)	1	2	3	4	5
Y	5	9	13	17	21

9) The total cost for a publishing company to publish a book ,is a fixed cost (100) plus a cost for each additional book that the company will print. Create a general relation between the number of printed books and the cost of printing.

X (# of Books)	0	100	200	300	400
Cost	100	300	500	700	900

10) Using the same data from problem 9, determine the cost of printing 543 books.

2.C. Understanding graphs of linear relationships

Remember, the ordered pairs (x,y) represent points in the system of axes. If we represent the points from the table below,

Point	X	Y
A	1	2
B	2	5
C	3	8
D	4	11

we will have the graph below.

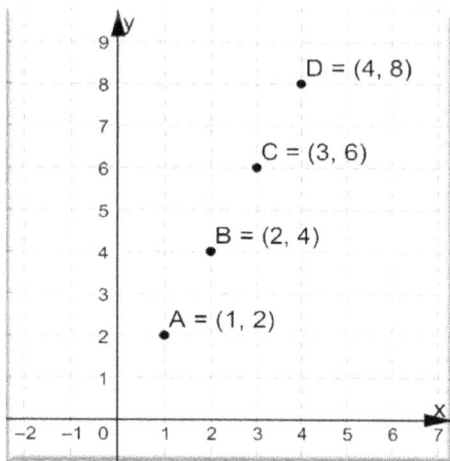

What we can see is that the points A, B, C, and D are situated on an imaginary straight line. If we connect these points, we will have a straight line.

Remember
- Straight line when the difference between successive values of x and y is the same respectively.
- When graphed the points are part of a line

EXAMPLE

Represent graphically the points from the table below and see if they are on a straight line.

Point	X	Y
A	1	1
B	2	3
C	3	5
D	4	7

PRACTICE

1) Represent graphically the points from the table below and see if they are part of a straight line.

Point	X	Y
A	1	2
B	2	4
C	3	6
D	4	8

2) Represent graphically the points from the table below and see if they are part of a straight line.

Point	X	Y
A	1	-4
B	2	0
C	3	4
D	4	8

3) Represent graphically the points from the table below and see if they are part of a straight line.

Point	X	Y
A	1	1
B	2	2
C	3	3

4) Represent graphically the points from the table below and see if they are part of a straight line.

Point	X	Y
A	1	-3
B	2	-1
C	3	1
D	4	3

5) Represent graphically the points from the table below and see if they are part of a straight line.

Point	X	Y
A	1	2
B	2	3
C	3	5
D	4	8

6) Represent graphically the points from the table below and see if they are part of a straight line.

Point	X	Y
A	1	2
B	2	4
C	4	5
D	5	8

7) Represent graphically the points from the table below and see if they are part of a straight line.

Point	X	Y
A	1	2
B	2	-1
C	3	6
D	4	-3

8) Represent graphically the points from the table below and see if they are part of a straight line.

Point	X	Y
A	1	1
B	2	3
C	3	5

9) Represent the points from the table below and see if they are part of a straight line.

Point	X	Y
A	-1	2
B	2	4
C	-3	6

10) Represent graphically the points from the table below and see if they are part of a straight line.

Point	X	Y
A	1	2
B	3	-3
C	5	6
D	4	7

2.D. The Distance between points

a. Horizontal distance

Let's suppose we have the points A (1,3) and B (5,3) As we can see the y coordinate is the same. If we represent these points on the cartesian system we get a horizontal segment that belongs to a horizontal line.

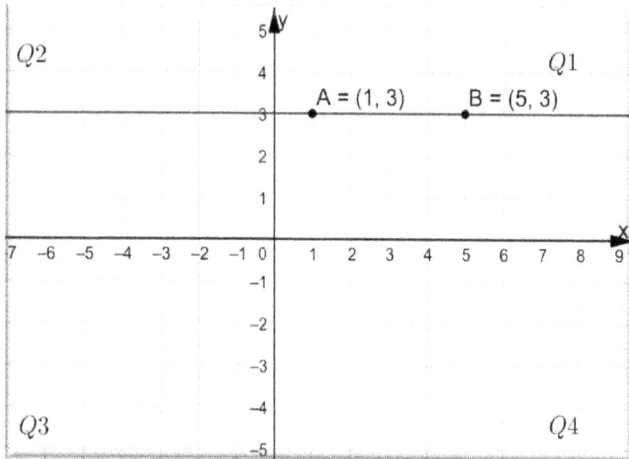

The distance between point A and point will be 5 – 1 = 4 units.

> **Remember:**
> The absolute value is always a positive number

Absolute value

Remember that the <u>absolute value</u> is always a positive number. We can think about the absolute value as the distance between the origin and that particular point, wherever the point is.

EXAMPLE

The absolute value of 7 is 7

The absolute value of -3.56 is 3.56

The notation for showing the absolute value is the number written between two vertical lines like below.

$|-7| = 7$ or $|-3.56| = 3.56$

If we go a bit more general, and consider Point A as point 1 and Point B as point 2, the coordinates of these two points will be written as:

A (x_1, y_1) and B (x_2, y_2)

The horizontal distance between A and B will be $|x_2 - x_1| = |5 - 1| = 4$

$|x_2 - x_1|$ represents the absolute value of the difference between the x coordinates of the points.

What happens when one of the x coordinates is negative?

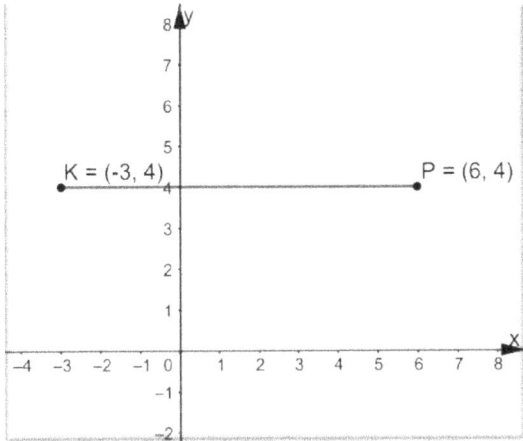

EXAMPLE

Let's calculate the horizontal distance between points K (-3,4) and P (6, 4)

K (x_1, y_1) and P (x_2, y_2)

The horizontal distance between K and P will be $|x_2 - x_1| = |6 - (-3)| = 9$

Remember:
All the points on a horizontal line have the SAME y coordinate

PRACTICE

Determine which answer is correct.

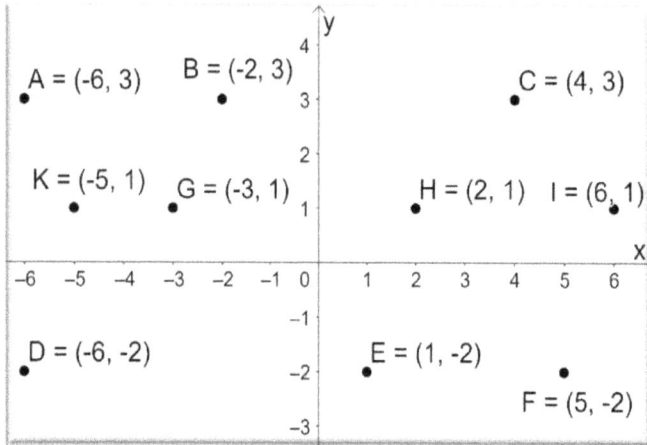

1) The distance between point A and point C is 10

2) The distance between point A and point B is 5

3) The distance between point D and point E is 7

4) The distance between point E and point F is 7

5) The distance between point K and point G is 2

6) The distance between point K and point H is 7

7) The distance between point K and point I is 9

8) The distance between point G and point H is 5

9) The distance between point G and point I is 8

10) The distance between point H and point I is 4

2.D. The Distance between points

b. Vertical distance

Let's suppose we have the points A (-3,1) and B (-3,6) As we can see the x coordinate is the same. If we represent these points on the cartesian system we get a vertical segment that belongs to a vertical line.

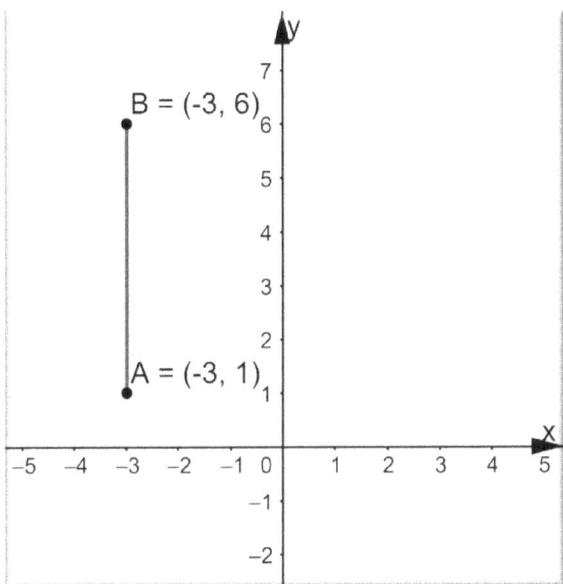

The vertical distance between point A and point B will be 6 − 1 = 5 units.

If we go a bit more general, and consider Point A as point 1 and Point B as point 2, the coordinates of these two points will be written as:

A (x_1, y_1) and B (x_2, y_2)

The vertical distance between A and B will be $|y_2 - y_1| = |6 - 1| = 5$

$|y_2 - y_1|$ represents the absolute value of the difference between the y coordinates of the points.

What happens when one of the y coordinates is negative?

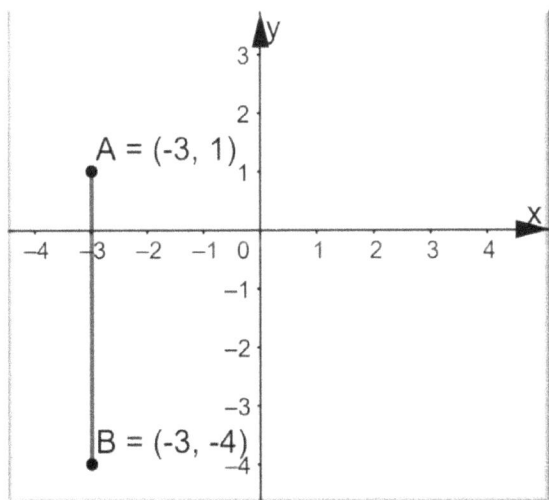

As we can see here, the y coordinates are 1 and -4 respectively.

The vertical distance between Point A and B will be:

$|y_2 - y_1| = |1 - (-4)| = 5$

Remember:
All the points on a
vertical line have the
SAME x coordinate

PRACTICE

Determine if the relations below are true.

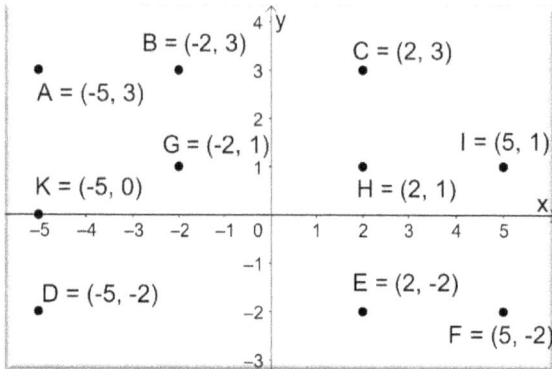

1) The distance between point A and point K is 10

2) The distance between point A and point D is 5

3) The distance between point B and point G is 3

4) The distance between point C and point H is 7

5) The distance between point C and point E is 5

6) The distance between point K and point D is 7

7) The distance between point H and point E is 3

8) The distance between point G and point E is 7

9) The distance between point A and point K is 9

10) The distance between point I and point H is 4

2.D. Distance between points

c. Non-horizontal and Non-vertical distance

What happens when the points are in a straight line tilted from down left to up right?
What is the distance in this case?

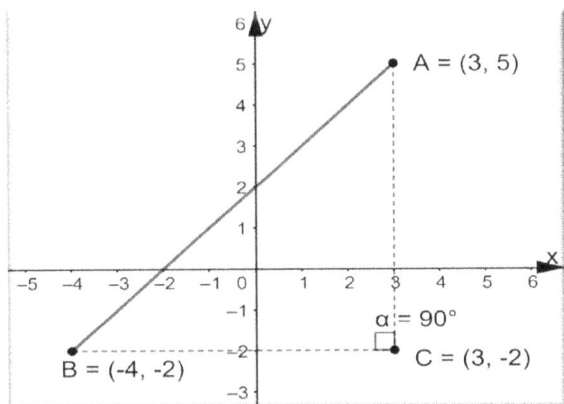

We would like to calculate the distance between point A and B. The easiest way is to create a right-angle triangle as follows:

> Remember:
> The distance between two non-vertical or non-horizontal points is calculated using the Pythagorean Theorem

We draw a horizontal line from B to C that will intersect the vertical line from A to C.

The angle in the triangle ABC is a right-angle (90^0).

In this triangle, the distance between A and B represents the hypotenuse.

As we remember, the Pythagorean theorem tells us that in a right-angle triangle, the hypotenuse squared equals the sum of one side squared plus the other side of the triangle squared.
Here we have:

$$AB^2 = BC^2 + AC^2$$

BC is a horizontal segment; so, distance BC = $|x_2 - x_1| = |3 - (-4)| = |7| = 7$
AC is a vertical segment: so, distance AC = $|y_2 - y_1| = |5 - (-2)| = |7| = 7$

Then AB = $\sqrt{BC^2 + AC^2} = \sqrt{7^2 + 7^2} = \sqrt{98} = 9.89$

PRACTICE

Determine which answer is correct.

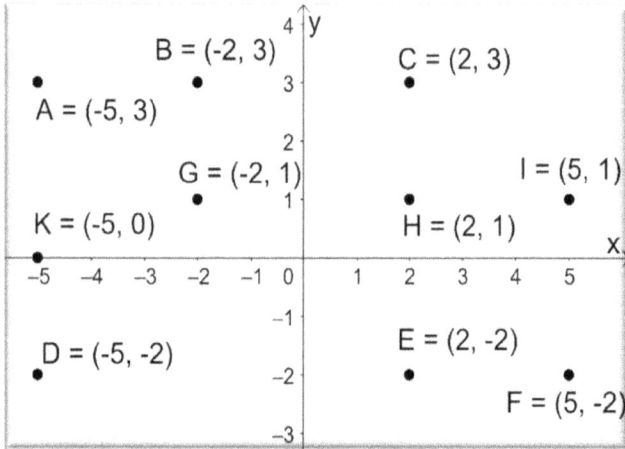

1) The distance between point B and point H is $2\sqrt{5} = 4.47$

2) The distance between point K and point C is 9.61

3) The distance between point A and point G is 3.6

4) The distance between point A and point H is 7.28

5) The distance between point A and point I is 20

6) The distance between point A and point F is 11.18

7) The distance between point A and point E is 8.6

8) The distance between point B and point D is 10

9) The distance between point B and point E is 6.4

10) The distance between point B and point F is 15

2.D. The Distance between points

d. Midpoint coordinates

If we have two points $A(x_1, y_1)$ and $B(x_2, y_2)$, we need to find the coordinates of the point $C(x, y)$ situated at the middle distance between A and B.

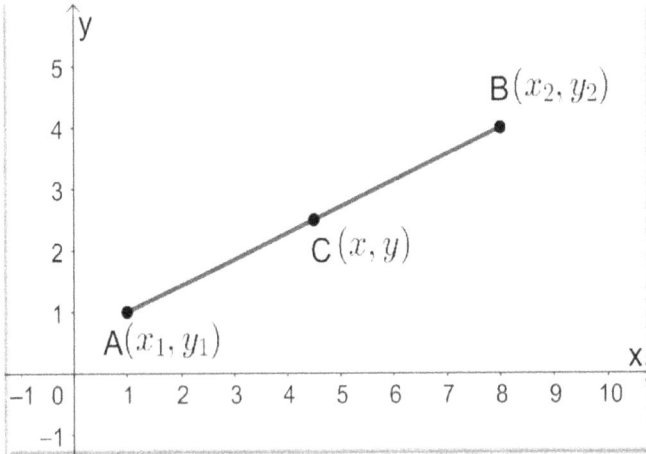

Because the distance between A and C equals the distance between C and B, to find the x coordinate, we can write:

$$x - x_1 = x_2 - x$$
$$2x = x_2 + x_1$$
$$x = \frac{x_2 + x_1}{2}$$

Remember:
The middle point coordinate between two other points is equal with the sum of the side coordinates divided by two

We do the same calculation on the y axis and have the formula of the y coordinate for the midpoint.

$$y - y_1 = y_2 - y$$
$$2y = y_2 + y_1$$
$$y = \frac{y_2 + y_1}{2}$$

EXAMPLE

Let's suppose the coordinates of A and B are shown in the figure below:

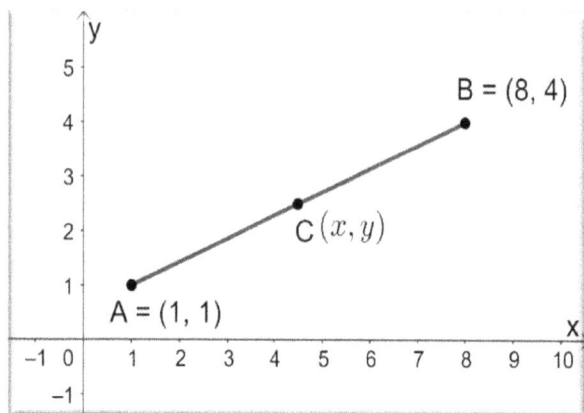

As we saw above, applying the formulas for the x and y coordinates of the midpoint C, we have:

$$x = \frac{x_2 + x_1}{2} = \frac{8+1}{2} = \frac{9}{2} = 4.5$$

$$y = \frac{y_2 + y_1}{2} = \frac{4+1}{2} = \frac{5}{2} = 2.5$$

So,

The midpoint coordinates are (4.5, 2.5

PRACTICE

Determine which answer is correct.

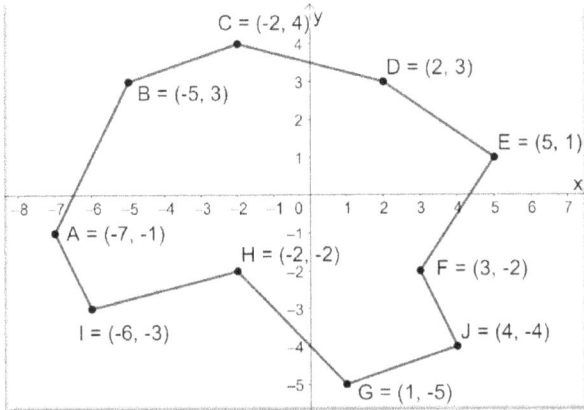

1) The mid-point coordinates of segment AB are: (-6,1)

2) The mid-point coordinates of segment BC are:(3,5)

3) The mid-point coordinates of segment CD are: (0,3.5)

4) The mid-point coordinates of segment DE are: (2,4)

5) The mid-point coordinates of segment EF are: (6,8)

6) The mid-point coordinates of segment FJ are: (3.5, -3)

7) The mid-point coordinates of segment JG are: (2.5, -4.5)

8) The mid-point coordinates of segment GH are: (0,-3)

9) The mid-point coordinates of segment HI are: (-4, -2.5)

10) The mid-point coordinates of segment IA are: (5, -3)

2.E. The Slope of a line

A slope of a line is the ratio between the "rise" of the line and the "run" of the line. The "rise" is the vertical distance or the difference between the y coordinates of any two points situated on the line

The "run" is the horizontal distance or the difference between the x coordinates of any two points situated on the line.

Let's suppose we want to find the slope of the line that passes through points P (x_1, y_1) and R (x_2, y_2).

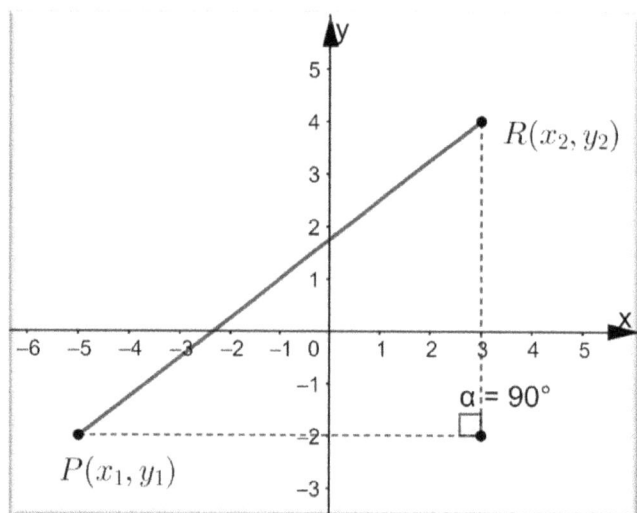

$$Slope = \frac{Rise}{Run} = \frac{Vertical\ distance}{Horizontal\ distance} = \frac{y_2 - y_1}{x_2 - x_1} =$$

m

We could say that this ratio shows how fast the vertical distance increases as the horizontal distance increases.

EXAMPLE

We have a line that passes through points A (3,4) and B (-5,-2)
Let's suppose that B is point 2 and A is point 1

$$Slope = m = \frac{Rise}{Run} = \frac{Vertical\ distance}{Horizontal\ distance} = \frac{y_2 - y_1}{x_2 - x_1} = \frac{-2 - 4}{-5 - 3} = \frac{-6}{-8} = \frac{6}{8} = \frac{3}{4}$$

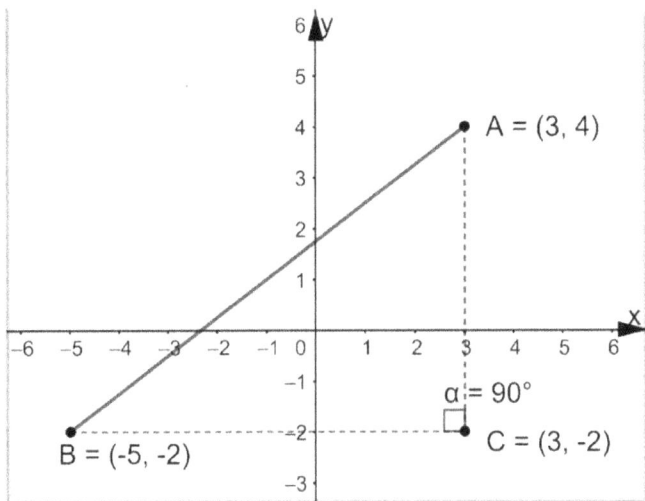

We could say that this ratio shows that the vertical distance increases by 3 units as the horizontal distance increases by 4 units.

Which will be the values of y as x increases from -5 to 3.

Using the relation for the slope that we just calculated, for x_2=-4 we get:

$$\frac{3}{4} = \frac{y_2-y_1}{x_2-x_1} = \frac{y_2-(-2)}{-4-(-5)} = \frac{y_2+2}{1}$$

So,

$$\frac{3}{4} = \frac{y_2+2}{1} \text{ using the cross-}$$

multiplication property

$3 = 4(y_2 + 2)$

$3 = 4y_2 + 8$

$3 - 8 = 4y_2$

$-5 = 4y_2$

$y_2 = \frac{-5}{4} = -1.25$

Remember:

The slope of a straight line is:

$$Slope = m = \frac{Rise}{Run} = \frac{Vertical\ distance}{Horizontal\ distance} = \frac{y_2 - y_1}{x_2 - x_1}$$

EXAMPLE

In the same way we can calculate the values for y for x increasing by 1 unit starting with   -5.

X	-5	-4	-3	-2	-1	0	1	2	3
Y	-2	-1.25	-0.5	0.25	1	1.75	2.5	3.25	4

As we can see, as the values of x increase, the values of y increase as well with the same ratio $\frac{3}{4}$.

Any increase of an unknown number or a variable (for example x) is symbolized Δx, where Δ is the Greek letter Delta.

So, the slope can be written as how fast the values of y vary as values of x vary.

$slope = \frac{\Delta y}{\Delta x}$ or could be thought as the <u>average rate of change</u>.

PRACTICE

Determine if the relations below are true.

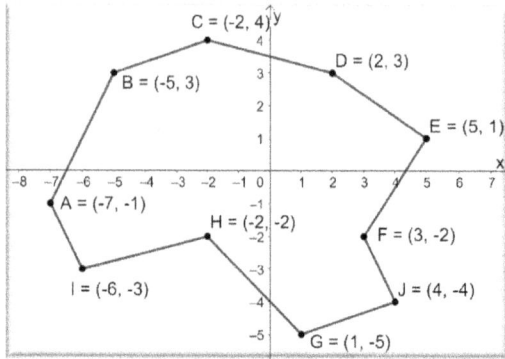

1) The slope of segment AB is: 2

2) The slope of segment BC is: 3

3) The slope of segment CD is: $-\frac{1}{4}$

4) The slope of segment DE is: 4

5) The slope of segment EF is: 5

6) The slope of segment FJ is: -2

7) The slope of segment JG is: $\frac{1}{3}$

8) The slope of segment GH is: 2

9) The slope of segment HI is: $\frac{1}{4}$

10) The slope of segment IA is: 7

Chapter 3

Linear Equations

3.A. Linear equations

What is an equation?

It is a mathematical "statement" which says that the <u>mathematical expression</u> to the left of the equal sign is exactly the same as the <u>mathematical expression</u> to the left of the equal sign.

The expressions we discussed before are linear equations. $x - 4 = 5$ is an equation.

X is called a variable or unknown. This variable has the exponent 1. This is one of the reasons we call these relations linear.

The word equation comes from "equating" or regarding the sides of the equation as equal.

EXAMPLE

Let's find the unknown we symbolize with letter x in the following equation:

$x - 3 = 4$

The question is what number should we use for x, so the left side of the equation equals the right side. In this case it is obvious that the number is 7. Here $x = 7$.

To find the unknown in an equation means to <u>solve</u> the equation.

> Remember:
> To <u>solve</u> an equation:
> To find the unknown in that equation.

3.B. Solving Linear Equations

a. Solve one-step linear equations:

First, we should talk about the balancing concept. What is this balancing concept? Whenever we have an equation, we could imagine that we have a balance that needs to be balanced all the time.

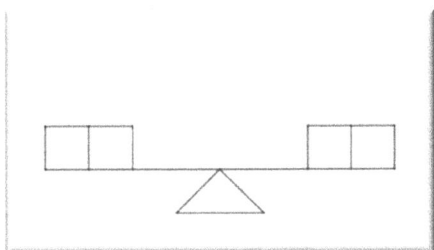

1. Let's suppose we have a balance. On each side we have two identical boxes. The balance is completely horizontal.

2. Now, we want to put the third identical box to the left side of the balance. We need to keep the balance horizontal. In order to keep the balance horizontal, we need to put another box to the right side of the balance.

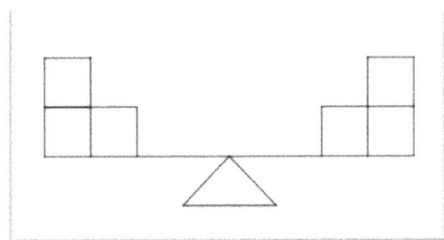

As we can see, we will add <u>exactly one</u> box on each side of the balance.

Think of the center of the balance as the equal sign of the equation. To keep this balance balanced or horizontal, we need to do the same action on each side of the balance.

3. Now, we want to take away two identical boxes from the left side of the balance. We need to keep the balance horizontal. In order to keep the balance horizontal, we need to take away another two identical boxes from the right side of the balance.

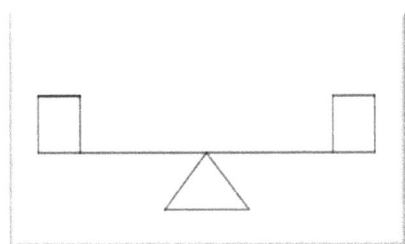

As we can see, we will take away <u>exactly two</u> boxes from each side of the balance.

> Remember:
> - to "isolate" x it means to have only x on one side of the equation, the rest of the expression should be on the other side.

EXAMPLE

1. We have the equation $x + 5 = 20$.

In order to <u>solve</u> the equation, or to find x, we need to <u>isolate</u> x.

- To do that we will need to cancel 5 from the left side.

- To cancel 5, we need to subtract the same 5 on the left side.

- To keep the equation balanced we have to subtract the same 5 on the right side.

3. To take away 5, we have;

$x + 5 = 20$

$\quad -5 \quad -5$

On the left side $+5 - 5 = 0$

Then,

$x + 0 = 20 - 5$

$x = 15$

> **Remember:**
> To check the solution –
> Substitute the solution in the original equation and check if the right side equals the left side of the equation.

We solved the equation by finding x.

To be sure that the result is correct, the next step is to <u>check</u> the solution.

EXAMPLE

We start with the previous equation:

$x + 5 = 20$

Where we found that $x = 15$

So,

We substitute 15 instead of x in the original equation.

The left side of the equation will be 15+5=20

The right side is equal with 20, so:

20=20

Now we can say for sure that $x = 15$ is the SOLUTION of the equation $x + 5 = 20$

PRACTICE

1) Solve and check the solution.

a) $x + 3 = 25$ 　　　　　　　　b) $x + \frac{1}{2} = 3\frac{1}{3}$

2) Write each sentence as an equation

a) x decreased by 3 is 7 　　　　　　　　b) The sum of 4 and x is 9

3) Check if the given value of x is a solution for the equation.

a) $x + 3 = 7$ for $x = 5$ 　　　　　　　　b) $-x - 2 = 7$ for $x = 4$

c) $x + 5 = 7$ for $x = 2$ 　　　　　　　　d) $x + 3.4 = 5.8$ for $x = 4.3$

4) What is the step needed to isolate the variable?

a) $x + 4 = 5$ b) $x - 3 = 15$

c) $4 = 7 + x$ d) $-4 = 6 + x$

5) Explain the error.

$x - \dfrac{2}{7} = \dfrac{5}{7}$

$x = \dfrac{3}{7}$

6) Write an equation. Solve and check.

Five more than n is negative seven

7) The length and the width of a rectangle add to 35 cm. If the width is 10 cm, what is the length?

8) A book price was reduced by $3.25 and then sold for $7.27. What was the original price of the book?

9) A right triangle has one of the angles (A) equal with 25 degrees. How big is the other angle beside the right one?

10) The 30cm Length of a rectangle is bigger that the Width by 12cm. How much is the Perimeter?

3.B. Solving Linear Equations

b. Solving two-step linear equations with addition and subtraction

Now we can go one step further and analyze the case when we have the unknown in both sides of the equation.

EXAMPLE

$3x - 5 = 2x + 2$

We will have to ISOLATE the unknown.

We will do the same operations in both terms so the equation remains balanced all the time.

Step 1: minus $2x$ in both sides of the equation.

$3x - 2x - 5 = 2x - 2x + 2$

$x - 5 = 2$

Step 2: add a 5 on both sides

$x - 5 + 5 = 2 + 5$

$x = 7$

> Remember:
> - to "isolate" x it means to have x only on one side of the equation, the rest of the expression should be on the other side.

To be sure the result is correct, we check by substituting $x = 7$ in the original equation.

CHECK

$3(7) - 5 = 2(7) + 2$

$21 - 5 = 14 + 2$

16=16

Indeed, $x = 7$ is the <u>solution</u> of the equation $3x - 5 = 2x + 2$

EXAMPLE

$4x + 3 = 3x - 5$

We will have to ISOLATE the unknown.

We will do the same operations in both terms so the equation remains balanced all the time.

Step 1: minus $3x$ in both sides of the equation.

$4x - 3x + 3 = 3x - 3x - 5$

$x + 3 = -5$

Step 2: subtract a 3 on both sides

$x + 3 - 3 = -5 - 3$

$x = -8$

To be sure the result is correct, we check by substituting $x = -8$ in the original equation.

CHECK

$4(-8) + 3 = 3(-8) - 5$

$-32 + 3 = -24 - 5$

$-29 = -29$

Indeed, $x = -8$ is the <u>solution</u> of the equation $4x + 3 = 3x - 5$

PRACTICE

1) Solve and check

$2x - 4 = x + 2$

2) Solve and check.

$5x + \frac{1}{3} = 4x + 1\frac{1}{2}$

3) Solve ad check

$3.25x - 5 = 2.25x - 6$

4) Check if these given values of x are the solution for the equation below.

$3x - 5 = 5x + 3$ $\qquad\qquad$ $x = 2, -3, -4$

5) Solve ad check

$2x + 4x - 4 = 5x - 3$ $\qquad\qquad$ We add the terms that have x (like 2apples+4apples)

6) Write an equation. Solve and check.

Four times a number increased by 3 is 3 times a number decreased by 1

$4x + 3 = 3x - 1$ $\qquad\qquad$ We subtract 3 on both sides.

7) Write an equation for which 3 is the solution

8) Isolate x

$7x - d = 6x + 2d$

9) Isolate x

$2x - a = x - 3a + b$

10) Ten times a number minus 2 is 9 times the number plus "d" plus 5. Write the equation, solve it for x and check it.

$10x - 2 = 9x + d + 5$

3.B. Solving Linear Equations

c. Solving two-step linear equations with multiplication and division

This type of equations appears when we have a constant like 2 multiplied with the variable (x) after we isolated the variable.

EXAMPLE

$4x + 3 = 8$

In this case, to isolate the variable x, we have to subtract 3 and then divide both sides of the equation with the same value 4.

Here we divide by 4

$\frac{4x}{4} = \frac{8}{4}$

$x = 2$

EXAMPLE

> Remember:
> The value we divide by is always the constant that multiplies the x variable

$4x + 2 = 8$

We will have to ISOLATE the unknown or the variable.

We will do the same operations in both terms so the equation remains balanced all the time.

Step 1: minus 2 on both sides of the equation.

$4x + 2 - 2 = 8 - 2$

$4x = 6$

Step 2: divide by 4 on both sides

$\frac{4x}{4} = \frac{6}{4}$

$x = \frac{6}{4} = \frac{3}{2}$

To be sure the result is correct, we check by substituting $x = \frac{3}{2}$ in the original equation.

CHECK

$4(\frac{3}{2}) + 2 = 8$

$\frac{4*3}{2} + 2 = 8$

$\frac{12}{2} + 2 = 8$

$6+2=8$

8=8

Indeed, $x = \frac{3}{2}$ is the <u>solution</u> of the equation $4x + 2 = 8$

PRACTICE

1) Solve: $5x - 2 = 4$

2) Solve and check $4x + 3 = 4 + x$

3) Solve $\frac{5}{x} = 3$ $x \neq 0$

4) Solve and check $3.2x + 5.2 = 9.4 - 2.1x$

5) Fifteen divided by a number is 5. Write then solve an equation to determine the number. Verify the solution.

6) Solve $-6x = 8 - 22x$

7) Solve and check $13 - 5x = 4 - 4x$

8) Two rental halls are considered for a private concert.
Hall A costs $70 per person
Hall B costs $2500, plus $35 per person
Determine the number of people for which the halls will cost the same to rent.

9) Seven subtract 4 times a number is equal to 5.3 times the same number, subtract 5. Write and solve the equation.

10) A car salesman is offered two methods of payment.
Plan A: $1400 per month with a commission of 5% on his sales.
Plan B: $1800 per month with a commission of 3% on his sales.
Sales are represented by the unknown x
a) Write an expression that will represent the total earnings using Plan A.
b) Write an expression that will represent the total earnings using Plan B.

c) Write an equation to determine the sales so the same total earnings are obtained from both plans.

d) Solve the equation and explain what the answer represents.

3.B. Solving Linear Equations

d. Solving two- step linear equations using distributive property

What is going to happen when we have the equation:

$2(x + 3) = 4$

We will first apply the distributivity property. We multiply the constant 2 with each of the terms in the bracket.

$2 * x + 2 * 3 = 4$

$2x + 6 = 4$

We subtract 6 from both sides.

$2x + 6 - 6 = 4 - 6$

$2x = -2$

We divide with 2 on each side.

$\frac{2x}{2} = \frac{-2}{2}$

$x = -1$

Check

We substitute $x = -1$ into the original equation to see if the right side equals the left side.

$2[(-1) + 3] = 4$

$2(2) = 4$

4=4

$x = -1$ is the solution of the equation $2(x + 3) = 4$

PRACTICE

1) Solve and check

$3(x + 2) = 7$

2) Solve and check

$5.2(x + 1.3) = 2.7$

3) Solve and check

$4(3x - 1) = 3(x + 5)$

4) Solve and check

$$\frac{1}{2}(3x - 4) = \frac{3}{2}(2x + 5)$$

5) Solve

$$\frac{3}{2}(1 + 4x) = \frac{1}{3}(3 - 2x)$$

6) Solve

$$\frac{x}{2} + \frac{x}{3} = x - \frac{1}{6}$$

7) Solve

$$\frac{x}{3} + \frac{5}{3} = \frac{3}{4}$$

8) Solve and check

$$3 - \frac{x}{12} = \frac{2x}{12} + 1$$

9) Solve

$$\frac{5}{18} - \frac{x}{18} = \frac{3x}{6} + \frac{1}{2}$$

10) Solve

$$\frac{1}{3}(2x - 3) + 4x - 3 = \frac{5}{6}(x + 1) + 2$$

3.C. Equation of a straight line

a. Non-vertical and non-horizontal line

We have seen that the slope of a straight line is calculated with the formula:

$$Slope = m = \frac{Rise}{Run} = \frac{Vertical\ distance}{Horizontal\ distance} = \frac{y_2 - y_1}{x_2 - x_1}$$

So, if we have a point on a straight-line $M(x_1, y_1)$, then the slope of the line that passes through point M and any other point $K(x, y)$ will be written as:

$$m = \frac{y - y_1}{x - x_1}$$

So, if we isolate y from here, we have:

$y - y_1 = m(x - x_1)$ **(Called point-slope form)**

$y = y_1 + mx - mx_1$

So,

$y = mx + y_1 - mx_1$

But, $y_1 - mx_1$ is a constant (b)

So, we have the equation of a straight line

$y = mx + b$ **(Called slope-intercept form)**

Where:

m – slope

b – intersection of the line with y axis. It is called <u>y-intercept</u>

> Remember:
> $y - y_1 = m(x - x_1)$ (Point-slope form)
> $y = mx + b$ (Slope-intercept form)
> $ax + bx + c = 0$ (General form)

EXAMPLE

Find the equation of a line that passes through point M (2,3) and has a slope m=3
We start with the point slope equation;

$$y - y_1 = m(x - x_1)$$

We plug the coordinates of point M and the slope.

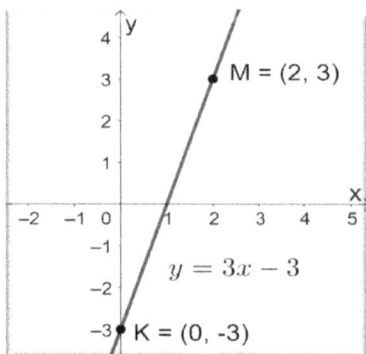

$y = 3x - 3$

$y - 3 = 3(x - 2)$

$y = 3 + 3x - 6$

$y = 3x - 3$

Here the slope is 3 and the intersection with the y axis is point K (0,-3)

Remember that the intersection of the line with x axis will have y coordinate equal to zero.

This is called x-intercept

$ax + bx + c = 0$ **(The general form of a linear equation)**

EXAMPLE

Represent the line $-8x + 4y - 4 = 0$

We transform the general form into the slope-intercept form in order to find the slope and the y intercept.

$-8x + 4y - 4 = 0$

Step 1: we add $8x$ on both sides of the equation.

$-8x + 8x + 4y - 4 = 0 + 8x$

$4y - 4 = 8x$

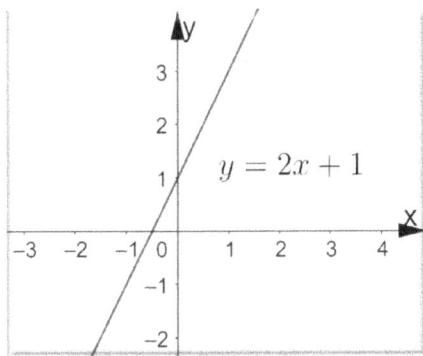

Step 2: we add 4 on each side of the equation.

$4y - 4 + 4 = 8x + 4$

$4y = 8x + 4$

Step 3: we divide by 4 on each side of the equation.

$\frac{4y}{4} = \frac{8x}{4} + \frac{4}{4}$

$y = 2x + 1$

From here we notice that the slope of the line is 2 and the intersection with y axis or y intercept is 1.

b) Y-intercept and x-intercept

y-intercept is the point where the straight line intersects the y axis.

What does it mean in terms of the coordinates?

The y-intercept is a point situated on y axis. This means that its x coordinate is zero.

EXAMPLE

Represent the points A(0,5); B(0,3); C(0,-2); D(0,-4) on a cartesian system.

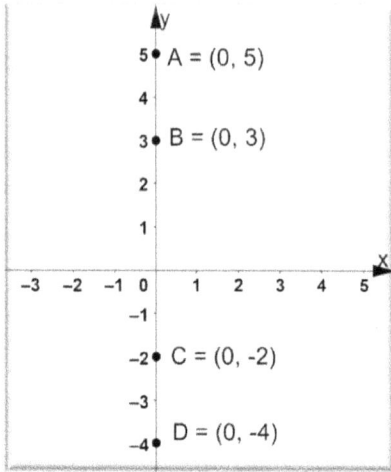

As we can see, all the points that are on the y axis have the x coordinate zero.

On the other hand, when we want to find the y intercept of a line, we will write the condition x=0

> Remember:
> y-intercept is the point where the straight line intersects the y axis.
> x-intercept is the point where the straight line intersects the x axis

x-intercept is the point where the straight line intersects the x axis.

What does it mean in terms of the coordinates?

The x-intercept is a point situated on x axis. This means that its y coordinate is zero

EXAMPLE

Represent the points A(4,0); B(3,0); C(-2,0); D(-3,0) on a cartesian system.

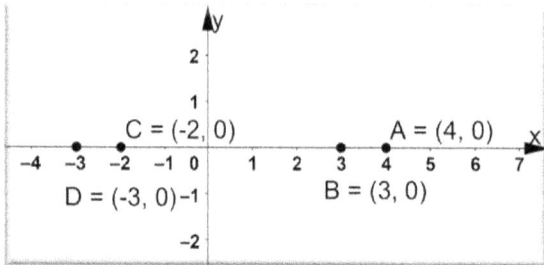

As we can see, all the points that are on the x axis have the y coordinate zero.

On the other hand, when we want to find the x intercept of a line, we will write the condition y=0

> Remember:
> To find the x intercept of a line, we will write the condition y=0

PRACTICE

Determine if the relations below are true.

1) The equation of the line through M (-3,1) and slope -2 is $y = -2x - 5$

2) The y intercept of the line $y = -2x - 5$ is y=-5

3) In the slope relation, $m = \frac{y-5}{x+4}$, the y intercept in terms of the slope m, is $b = 6m + 4$

4) The equation of the parallel line with $y = 3x + 1$ that passes through the point M $(5,6)$ is $y = 2x - 9$

5) Y intercept of the parallel line with $y = 3x + 1$ in problem 4 is b=-9

6) The intersection to x axis of $y = 4x - 8$ is P $(3,0)$

7) The intersection to x axis of the line $y = -3x + 1$ is $P(\frac{1}{3}, 0)$

8) The slope of the line with x intercept = 4 and passing through M(2,3) is $m = -1$

9) The slope of the line that has as x intercept the point P(2,0) and passes through the point M(6,1. is m = -1/4

10) The slope of line BC is $m = 4$

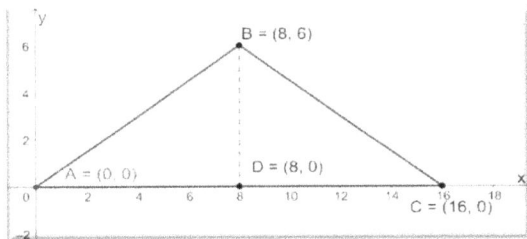

11) Determine the slope of each line
a) $y = 3x - 4$ b) $y = -2x + 3$ c) $y = 4x - 7$

12) Graph the line $2(x - 3) = (y + 1)$

13) Determine the slope and graph the line: $3x + 4y - 6 = 0$

14) Plot the following equations on the same graph. What do you notice?
a) $5x + 3y - 2 = 0$ b) $3x + 3y - 2 = 0$ c) $x + 3y - 2 = 0$

3.D. Straight-line graph

a. From equation to the graph

We have seen that the slope of a straight line is calculated with the formula:

$$Slope = m = \frac{Rise}{Run} = \frac{Vertical\ distance}{Horizontal\ distance} = \frac{y_2 - y_1}{x_2 - x_1}$$

So, if we have a point on a straight-line $M(x_1, y_1)$, then the slope of the line that passes through point M and any other point $K(x, y)$ will be written as:

$$m = \frac{y - y_1}{x - x_1}$$

So, if we isolate y from here, we have:

$$y = mx + b$$

Where:

m – slope

b – intersection of the line with y axis.

It is called <u>y-intercept</u>

> **Remember:**
> To graph a straight line, the form should be:
> $$y = mx + b$$
> Where:
> m – slope
> b – intersection of the line with y axis.

EXAMPLE

Graph the straight-line represented by $y = 2x + 3$

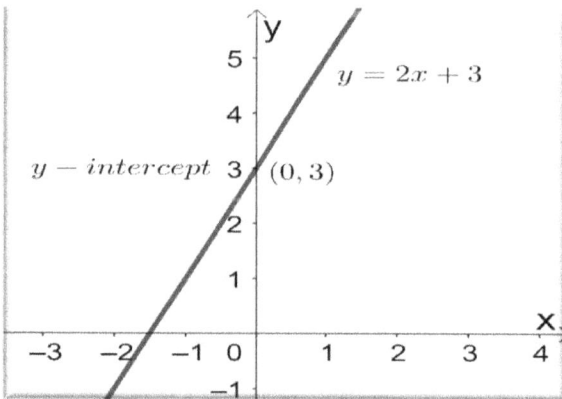

Here, the slope is 2, and the y intercept is the point (0,3)

EXAMPLE

Analyze each of these equations of straight lines and graph them.

a) $y = \frac{1}{2}x$

b) $y = -3x + 2$

c) $y = -x - 1$

d) $y = 3$

e) $x = -2$

Let's analyze them separately.

a) $y = \frac{1}{2}x$

the slope here is 0.5 and the y intercept is (0,0)

b) $y = -3x + 2$

Whenever the slope is negative, the line is tilted from down right to up left.

Here, the slope is -3 and the y intercept is (0,2).

c) $y = -x - 1$

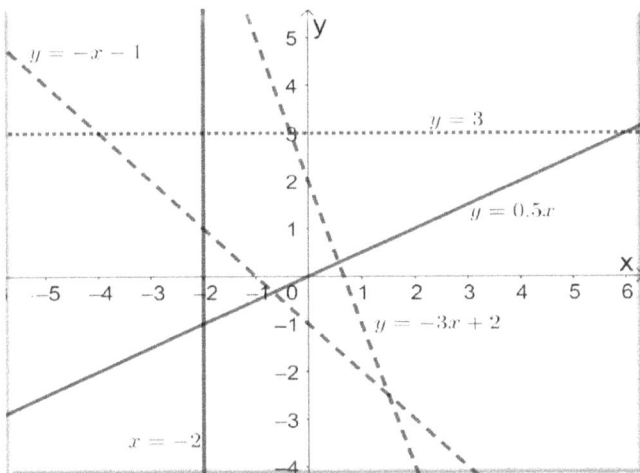

Here, the slope is negative (-1), so the line is tilted from down right to up left. The y intercept is (0,-1)

d) $y = 3$

Here, there is no x. this line is a horizontal line that goes through point (0,3) and is parallel with x axis.

e) $x = -2$

This is a vertical line that that goes through point (-2,0) and is parallel with y axis.

b. From graph to the equation

Next, is the case when we are given the line, and we have to determine the mathematical equation of the line.

EXAMPLE

Find the equations of lines a) and b).

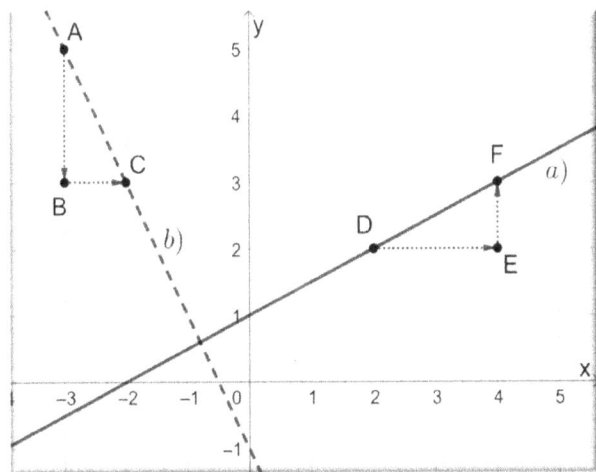

Line a)

We notice that y intercept is 1, so the constant in the equation is known.

To find the slope we start at point D, go horizontally (run) for +2 units to the right, then go vertically (rise) up plus 1 unit. So, the slope is calculated as:

$$slope = \frac{rise}{run} = \frac{1}{2}$$

The equation is $y = \frac{1}{2}x + 1$

Line b)

We notice that y intercept is -1, so the constant in the equation is known.

To find the slope we start at point A, go vertically ("rise" downward) for -2 units, then go horizontally (run) plus 1 unit. So, the slope is calculated as:

$$slope = \frac{rise}{run} = \frac{-2}{1} = -2$$

The equation is $y = -2x - 1$

PRACTICE

Determine which answer is correct.

1) The slope of the line $y = 3x - 1$ is 3

2) The y intercept of the line $2x + 3y = 5$ is $\frac{25}{3}$

3) The slope of the line $-3x + 5y = -1$ is $-\frac{1}{5}$

4) The y intercept of the line $2(x - 5) + 3(y + 2) = 2$ is y=52

5) The slope of the line $-3(x + 2) - 4(y - 7) = 6$ is $-\frac{3}{4}$

The next problems use the figure shown below

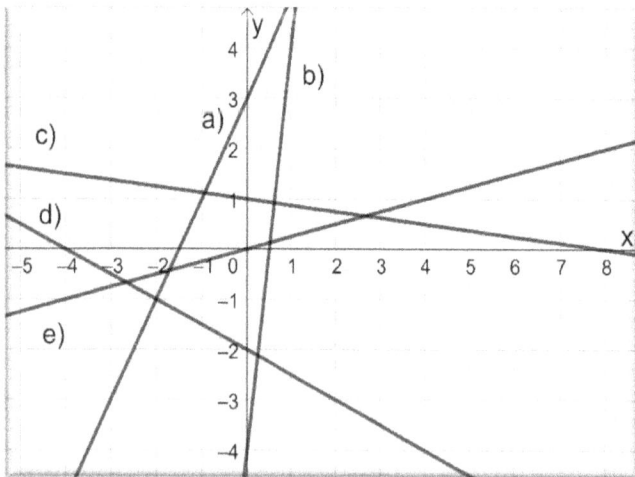

6) Line a) has the equation
Y=2x+33

7) Line b) has the equation
Y= 8x-49

8) Line c) has the equation
$Y = -\frac{1}{8}x + 1$

9) Line d) has the equation
$y = \frac{1}{2}x - 2$

10) Line e) has the equation
$y = \frac{1}{4}x$

3.E. Special cases of linear equations:

Vertical and horizontal lines

We have seen in____???_____ that the slope of a straight line is calculated with the formula:

$$Slope = m = \frac{Rise}{Run} = \frac{Vertical\ distance}{Horizontal\ distance} = \frac{y_2 - y_1}{x_2 - x_1}$$

So, if we have a point on a straight-line $M(x_1, y_1)$, then the slope of the line that passes through point M and any other point $K(x, y)$ will be written as:

$$m = \frac{y - y_1}{x - x_1}$$

Vertical line

Let's see an example of a vertical line. If we represent the coordinates of a few points on this vertical line, say x=3, we will notice that each point on this vertical line has the same x coordinate. This x coordinate is always 3. In this case the equation of a vertical line that passes through x equal 3 will have the equation

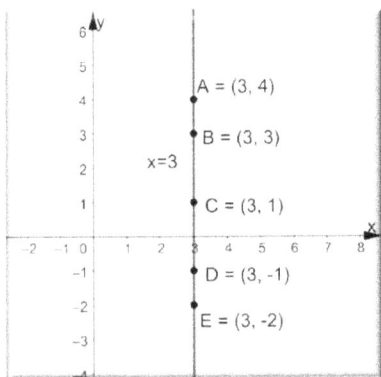

$$x = 3$$

> Remember:
> - The equation for a vertical line is $x = constant$
> - The slope m is infinite, or undefined

We see that the equation for a vertical line is $x = constant$

This means that for any values of y in the formula $m = \frac{y - y_1}{x - x_1}$, the x coordinates will always be the same. But, if the x is the same, the denominator of the slope formula will become zero. For the constant x, the slope m is infinite, or undefined.

Horizontal line

Let's see an example of a horizontal line. If we represent the coordinates of a few points on this horizontal line, say y=-2, we will notice that each point on this line has the same y coordinate. This coordinate is always minus 2. In this case the equation of a vertical line that passes through y equal minus 3 will have the equation

$$y = -2$$

The equation for the horizontal line is $y = constant$

In this case, the y coordinates of any point on the horizontal line are the same. From here we have that the difference $y - y_1$ will be zero.

For the constant y, the slope m is equal to zero.

EXAMPLE

Find the slope of the line with the equation y=3

Let's consider point one N (5,3)

and point two M (9,3) situated on the line.

Let's calculate the slope.

$m = \dfrac{y_2 - y_1}{x_2 - x_1} = \dfrac{3-3}{9-5} = \dfrac{0}{4} = 0$

Remember:
- The equation for the horizontal line is $y = constant$
- The slope m is equal to zero

PRACTICE

Determine if the relations below are true.

1) The equation of a horizontal line passing through M (3,4) is $y = 4$

Problems 2,3,4,5 will be based on the figure shown below.

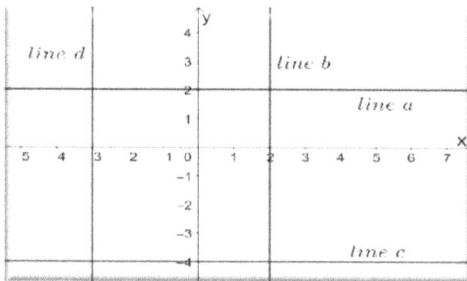

2) The equation of line a is $y = 2$

3) The equation of line b is $x = 3$

4) The equation of line c is $y = -4$

5) The equation of line d is $x = +3$

Problems 6 and 7 will be based on the figure shown below.

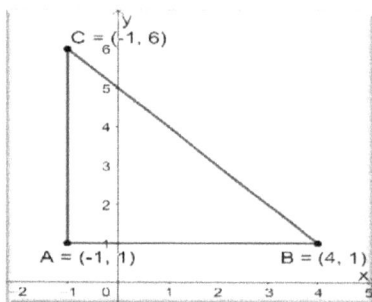

6) The equation of the line that passes through the points C and A in the figure below is $x = -1$

7) The equation of the line that passes through the points A and B in the figure below is $x = -8$

Problems 8 and 9 will be based on the figure shown below.

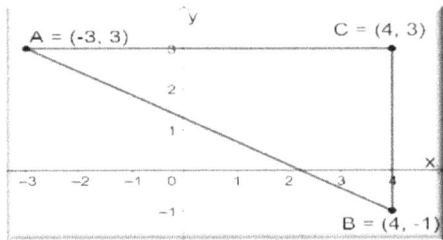

8) The equation of segment AC is $y = 1$

9) The equation of segment CB is $x = 4$

Problem 10 will be based on the figure shown below.

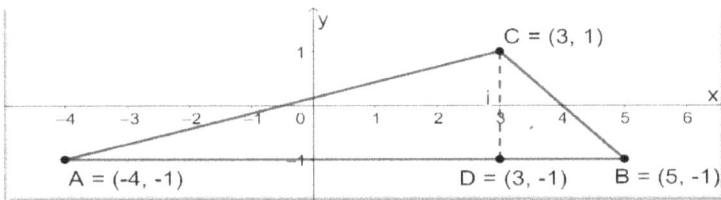

10) The equation of segment CD is $x = -2$

3.F. Parallel and perpendicular lines

a. Parallel lines

Remember that we have the formula for a straight line as:

$y = mx + b$

In this context, two lines are parallel when they have the same slope. The y intercept has to be different.

EXAMPLE

Graph the following straight lines.

$y = 2x + 1$

$y = 2x + 4$

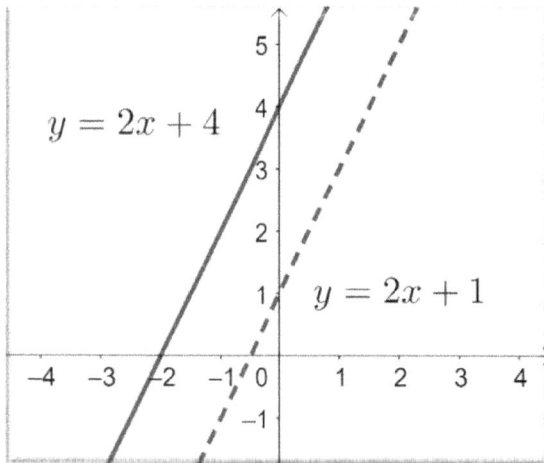

As we can see in the graph, the equations have the same slope, in this case 2, and different y intercepts, in this case 1 and 4 respectively.

Question:

What happens when the slopes are the same, and the y intercepts are the same as well?

Remember:
Two parallel lines have the same slope

EXAMPLE

We have the lines.

$y_1 = 2x + 1$

$y_2 = 2x + 1$

We can see that these lines y_1 and y_2 are identical. Indeed, if we represent them on the cartesian system, it will be exactly the same line. We will have only one line.

PRACTICE

Determine if the relations below are true.

1) The equation of a line parallel with $y = x - 1$ that intersects y axis at point M (0,5) is $y = x + 5$

2) The equation of a line parallel with $y = -3x + 2$ that intersects y axis at point M (0,-3) is $y = 3x + 3$

3) The line $y = -5x + 3$ is not parallel with $y = -4x + 3$

4) The $y = 3x - 1$ is the same as $y = 3x + 1$

5) The line $y = 4x + 3$ is parallel with $y = x - 25$

Problems 6 to 10 will be based on the figure shown below.

6) Line a is parallel to line b

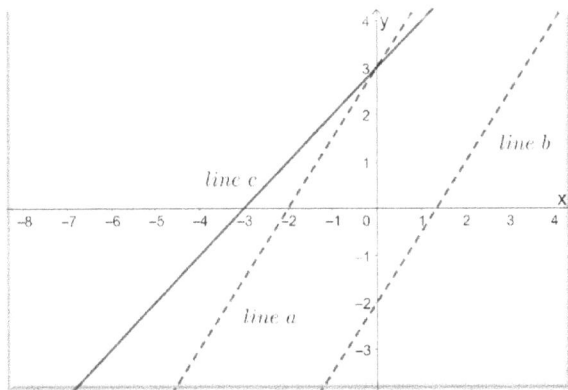

7) The equation of the line a is $y = \frac{2}{3}x + 3$

8) The equation of the line b is $y = \frac{3}{2}x - 2$

9) Line c is parallel with line a and line b

10) The equation of the line c is $= x + 3$

3.F. Parallel and perpendicular lines

b. Perpendicular lines

Remember that two lines that are not vertical are perpendiculars (i.e. have an angle of 90^0 between them) if the product of their slopes is negative 1.

This means that, if the first line's slope is m_1, and the second line's slope is m_2, then these lines are perpendicular if $m_1 \times m_2 = -1$, or $m_1 = \frac{-1}{m_2}$

EXAMPLE

We have the line given by the equation:

$y_1 = 2x + 1$

Show that the line given by the equation:

$y_2 = -0.5x + 4$

is perpendicular to the first line.

So,

$m_1 = 2$, and $m_2 = -0.5$

Then,

$m_1 \times m_2 = 2 \times (-0.5) = -1$

From here it results that y_1 and y_2 are perpendiculars.

These two lines are represented in the graph below.

> Remember:
> Lines are perpendicular if $m_1 \times m_2 = -1$, or $m_1 = \frac{-1}{m_2}$
> The product between the slopes of the two lines is equal with -1

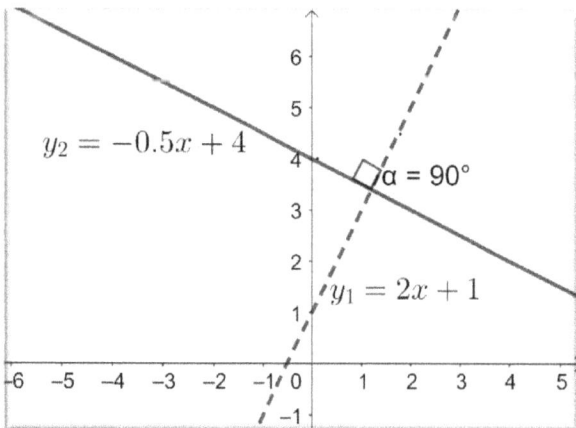

$y_2 = -0.5x + 4$

$\alpha = 90°$

$y_1 = 2x + 1$

PRACTICE

Determine if the relations below are true.

1) The lines that have the equations $y = 3x + 4$ and $y = -\frac{1}{3}x - 5$ are perpendicular.

2) The lines that have the equations $y = 3x - 3$ and $y = 3x + 3$ are not perpendicular.

3) The line perpendicular to the line $y = -2x + 7$ has the slope $m = 2$.

4) The equation of the line perpendicular to $y = 5x - 1$ that passes through M (3,4) is $y = -\frac{1}{5}x + 4.6$

5) The equation of the line perpendicular to $y = -2x + 3$ through M (-2,-3) is $y = \frac{1}{2}x + 2$

6) The equation of the line perpendicular to $y = x + 2$ through K (1,2) intersects the y axis in M (0,3)

7) The equation of the line perpendicular to AB in point B, is $y = \frac{1}{3}x + 10$

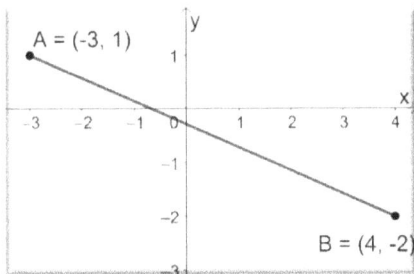
A = (-3, 1)
B = (4, -2)

8) The equation of the line perpendicular to AB in problem 7 will intersect x axis in $x = 1$

9) The perpendicular to AB through the point B (6,1) will intersect y axis in $b = 13$

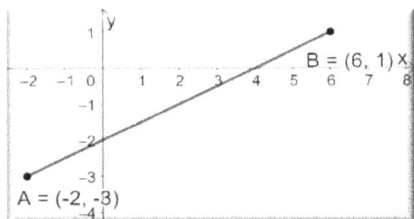
B = (6, 1)
A = (-2, -3)

10) The perpendicular to AB through the point B (6,1) will intersect x axis at $x = 2$

Examples of applications of linear relationships

Rate of Change: Real World Application

EXAMPLE

A high school student begins his practice for the next race. At 4:00 pm he starts to run. At 5:30 pm, the student finishes the run. He has run a total of 6.5 miles. How fast was his average speed over the course of the run?

We know that the rate of change is in fact the speed of his run or, distance over time. The two variables here are time on x axis and distance on y axis. We choose the first point at 4:00 pm. This is the beginning time so let's set y to zero. So, our first point on the time-distance graph is (0,0). Our time is in hours. Our second point is 1.5 hours later (on x axis), and, because the student ran 8.5 km, the second point on the graph will be: (1.5,8.5). Our speed (rate of change) is simply the slope of the line connecting these two points. The slope, given by:

$$m = \frac{y_2 - y_1}{x_2 - x_1} = \frac{8.5 - 0}{1.5 - 0} = \frac{8.5}{1.5} = 5.66 \; km/h$$

EXAMPLE

Graph the line illustrating speed

To graph this line, we need the y-intercept and the slope to write the equation. The slope was 5.66 km per hour and since the starting point was at (0,0), the y-intercept is 0. So, our final relation is $y = 5.66x$

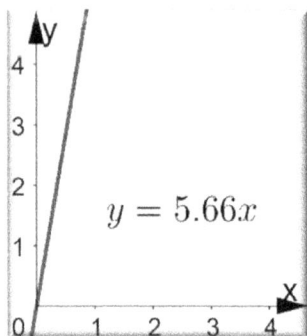

The time-distance graph.

The line has the equation $y = 5.66x$. The two variables are time on x axis and distance on y axis. Using this graph, we could make predictions, assuming that the average speed remains the same.

With this new function, we can now answer some more questions.

- How many miles did the student run after the first hour? Using the equation, time (x) is 1 and the distance would be y=5.66*1=5.66 km.
- If he kept running at the same pace for a total of 4 hours, how many km will he have run? The time (x) is 4 hours so the distance y=5.66*4=22.64km.

There are many such applications for linear equations. Anything that involves a constant rate of change can be nicely represented with a line with the slope.

EXAMPLE

Renting a RV

A rental company charges a flat fee of $20 and an additional $0.75 per km to rent a RV. Write a linear equation to approximate the cost (in dollars on y axis) in terms of the number of km driven on x axis. How much would a 88 km trip cost?

Using the slope-intercept form of a linear equation, with the total cost labeled y or dependent variable and the km labeled x or independent variable:

$y = mx + b$

The total cost is equal to the rate per km times the number of km driven plus the cost for the flat fee:

$y = 0.75x + 20$

To calculate the cost of an 88-km trip, we substitute 88 for x into the equation:

$Cost = 0.75 * 88 + 20 = 66 + 20 = \$86$

Chapter 4

Linear Inequalities

4.A. Express linear inequalities graphically and algebraically

Let's start with a x value of, say 5. We represent it on the number line as a point at

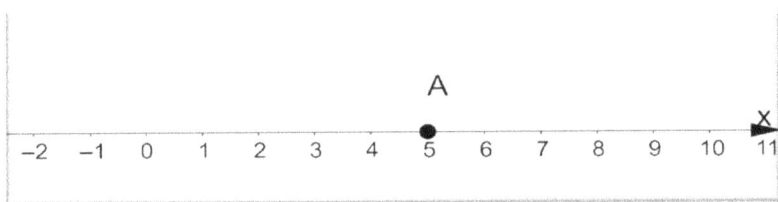

five units to the right of zero, or 5 units to the right of zero.

Algebraically, we write $x = 5$

When instead, we refer to all the numbers smaller than, say 5, including 5 we represent this case as an arrow above the number line towards left, starting with a filled circle.

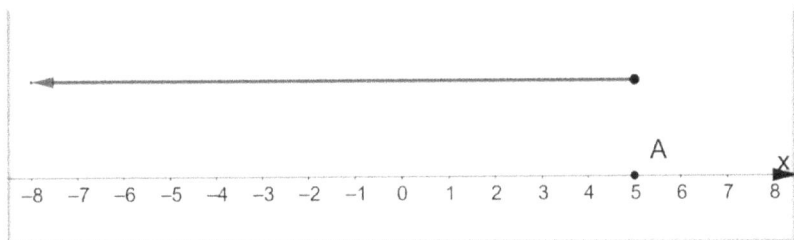

Algebraically we write: $x \leq 5$

When we refer to all the numbers smaller than, say 3, excluding 3, we represent an arrow above the number line towards left, starting with an empty circle.

Algebraically we write: $x < 3$

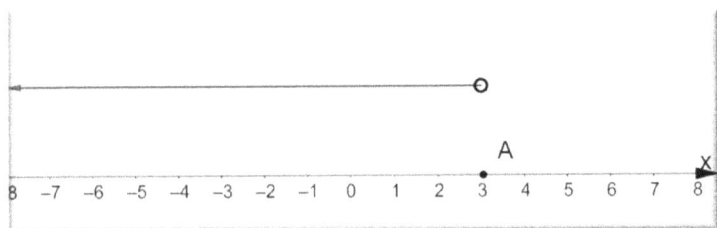

When, we refer to all the numbers bigger than, say -3, including -3, we represent this

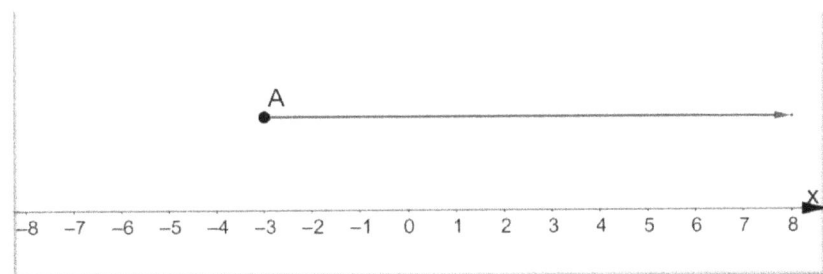

case as an arrow above the number line towards right, starting with a filled circle.

Algebraically we write:

$x \geq -3$

When we refer to all the numbers bigger than, say -7, excluding -7, we represent an

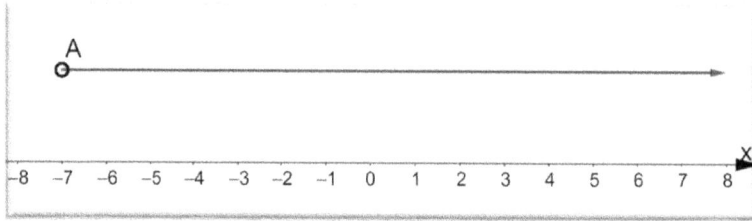

arrow above the number line towards right, starting with an empty circle.

Algebraically we write:

$x > -7$

EXAMPLE

Represent on the line and algebraically:

a. a number bigger or equal to 0

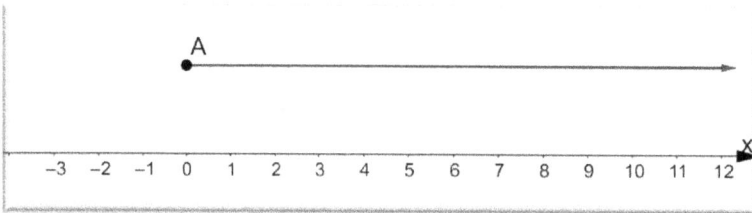

Algebraically we write:

$x \geq 0$

b. all numbers smaller and equal to 4

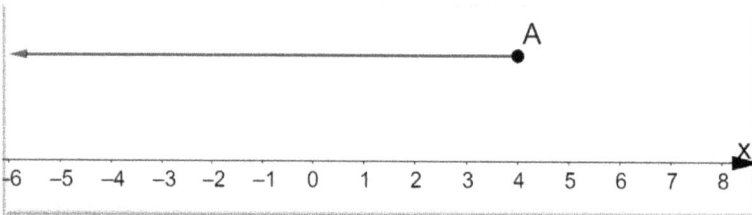

Algebraically we write:

$x \leq 4$

PRACTICE

1) Represent on the number line and algebraically:
A number bigger than and equal to -5

2) Represent on the number line and algebraically:
A number bigger than and equal to 1

3) Represent on the number line and algebraically:
A number less than and equal to 2

Remember:
$\leq$ is represented by an arrow to the left starting with a full circle.
$<$ is represented by an arrow to the left starting with an empty circle.
$\geq$ is represented by an arrow to the right starting with a full circle.
$>$ is represented by an arrow to the right starting with an empty circle.

4) Is each inequality true or false?

a) $3 < 7$
b) $5 \leq -8$
c) $\frac{1}{5} < \frac{1}{7}$

5) Use a symbol $>, <, \leq, or \geq$ to write an inequality that corresponds to each statement.

a) x is less than -7

b) a number is greater and equal than 3

c) x is negative

6) Is each number a solution of $x \leq -3$?

a) 0
b) 5
c) -7

7) Write 3 numbers that are solutions of each inequality.

a) $a > 3$
b) $b \leq 3$
c) $w < -5$

8) Determine whether the given number is a solution.

a) $y < 2, 2$
b) $x > 4, 0$
c) $z \leq 5, 0$

9) Iveta and Emma write the inequality whose solution is shown below.

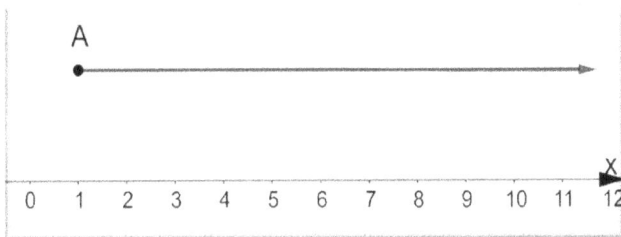

Iveta writes $x \geq 1$

Emma writes $1 > x$

Who is correct?

10) Graph the solution for $x \leq -2$

4.B. Solving one-step linear inequalities

Let's have the inequality:

$x + 3 \leq 7$

Like in case of equations the idea is to isolate the unknown.

We minus 3 in both sides of the inequality.

$x + 3 - 3 \leq 7 - 3$

So,

$x \leq 4$

According to 3.1 we can represent this on number line.

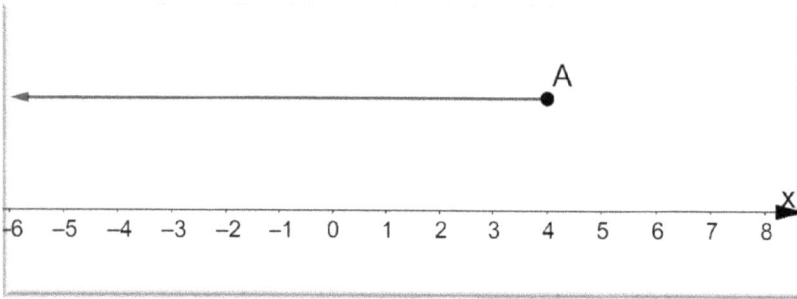

Let's have the inequality:

$x - 5 > 1$

Like in case of equations the idea is to isolate the unknown.

We add 5 in both sides of the inequality.

$x - 5 + 5 > 1 + 5$

So,

$x > 6$

According to 3.1, we can represent this solution on the number line.

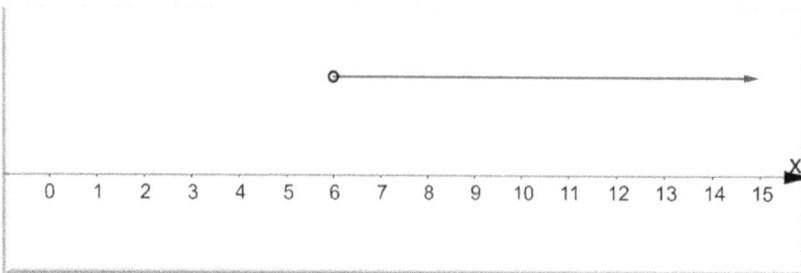

What would happen when we have:

$-x + 4 > 12$

We subtract 4 in both sides.

$-x + 4 - 4 > 12 - 4$

$-x > 8$

We want x not $-x$ so, we multiply with minus one on both sides of the inequality.

In this case, the sign of the inequality changes from $>$ to $<$

$x < -8$

EXAMPLE

Solve for x

$2x \geq 6$

We divide both sides with the coefficient of x.

$\frac{2x}{2} \geq \frac{6}{2}$

$x \geq 3$

> Remember:
> When we multiply or divide by -1, the inequality changes the sign from what it is to the opposite.

PRACTICE

1) Solve and graph the solution on the number line.

$x - 7 \geq -4$

2) Which operation will you perform on each side of the inequality to isolate the variable?

a) $y + 3 > 5$ b) $k - 3.4 < 7.8$ c) $5.6 < 2.1 + n$

3) What must be done to the first inequality to get to the second inequality?

a) $x - 5 \leq 4$ b) $2x \leq 4$ c) $x - \frac{1}{3} > \frac{2}{3}$

$x \leq 9$ $x \leq 2$ $x > 1$

4) State three values that satisfy each inequality; one integer, one fraction, and one decimal

a) $x + 3 < 5$

b) $x - 4 > 1$

c) $5x \leq 20$

5) Solve, graph and check the inequality:

$4x \leq 12$

6) Is 2 a solution of the inequalities below?

a) $x - 3 > -4$ 　　　　　　　　　 b) $y + 4 \leq 5$ 　　　　　　　　 c) $m + 2 < 6$

7) Melissa has \$310 in her bank account. She must maintain a minimum balance of \$600 in her bank account to avoid paying a monthly fee. How much money can Melissa deposit into her account to avoid paying bank fees?

a) Choose a variable and write an inequality to solve the problem.

b) Solve the problem

8) Mark is saving money to buy a camera for the next camping trip. He earned \$200 during the weekends, but he still did not have the \$750 he needed for the camera.

a) Choose a variable for the money needed, then write an inequality to represent this situation.

b) Solve the inequality.

c) Verify the solution.

9) Write and solve an inequality to show how many cars Samuel has to wash at \$7 a car to earn at least \$350?

10) A water slide charges \$2 to rent an inflatable ring, and \$0.5 per ride. Iveta has \$12. How many rides can Ivcta go on?

4.C. Solving multi-step linear inequalities

Let's suppose we have:

$3x - 5 > 10$

First, we add 5 to each side to cancel -5 in the left side.

$3x - 5 + 5 > 10 + 5$

$3x > 15$

Second, we divide with the coefficient of the unknown.

$\frac{3x}{3} > \frac{15}{3}$

We get:

$x > 5$

EXAMPLE

$2x + 4x - 5 > 10 + 3$

Step 1: add 5 in both sides to cancel 5 in the left side.

$2x + 4x - 5 + 5 > 10 + 3 + 5$

Step 2: deal with the like terms in both sides.

$6x > 18$

Step 3: divide with the coefficient of the unknown in both sides.

$\frac{6x}{6} > \frac{18}{6}$

$x > 3$

EXAMPLE

$5x - 7 > 3x - 18 + 3$

Step 1: add 7 in both sides to cancel 7 in the left side.

$5x - 7 + 7 > 3x - 18 + 3 + 7$

$5x > 3x - 6$

Step 2: subtract $3x$ in both sides to isolate the unknown on the left side.

$5x - 3x > 3x - 3x - 8$

$2x > -8$

Step 3: divide with the coefficient of the unknown in both sides.

$\frac{2x}{2} > -\frac{8}{2}$

$x > 4$

EXAMPLE

$\frac{(x+2)}{2} < \frac{(3x-5)}{3}$

Step 1: multiply both sides with the common denominator 6

$\frac{6*(x+2)}{2} < \frac{6*(3x-5)}{3}$

$3 * (x + 2) < 2 * (3x - 5)$

$3x + 6 < 6x - 10$

Step 2: subtract 6 in both sides

$3x + 6 - 6 < 6x - 10 - 6$

$3x < 6x - 16$

Step 3: subtract $6x$ in both sides.

$3x - 6x < 6x - 6x - 16$

$-3x < -16$

Step 4: multiply with minus 1 in both sides.

The sign of the inequality will flip from < to >

$3x > 16$

Step 5: divide with the coefficient of the unknown.

$\frac{3x}{3} > \frac{16}{3}$

$x > 5\frac{1}{3}$

EXAMPLE

$\frac{1}{3}x + 3 > 9$

$\frac{1}{3}x + 3 - 3 > 9 - 3$

$\frac{1}{3}x > 6$

Multiply with 3 to cancel 1/3

$\frac{3*1}{3}x > 6 * 3$

$x > 9$

PRACTICE

1) Solve and check

$5x + 7 \geq 2$

2) Solve and graph the solution

$3x + 4 \geq 6 + 2x$

3) Solve

$6.2x + 3.1 < 2.3x + 1.3$

4) Solve and check

$1 + x > 3 + \frac{1}{3}x$

5) Solve

$\frac{2}{5}x - \frac{1}{2} > 3 + x$

6) Your school wants to raise money for charity. The school organizes a dance where the DJ costs $1200 and the ticket costs $8. How many tickets have to be sold to make a profit more than $1700?

a) Write an inequality to solve the problem

b) Solve and verify the solution

7) Solve

$7 + \frac{1}{3}x > 2(x + 12)$

8) Solve

$3(x - 3) > \frac{2}{3}(3x + 6)$

9) Solve

$\frac{1}{2}x + \frac{5}{3} \leq \frac{3}{2}x - \frac{1}{4}$

10) John is replacing the light bulbs in his house from regular to energy saver light bulbs.

A regular light bulb costs $0.6 and has an electricity cost of $0.005 per hour.

An energy saver light bulb costs $5.5 and has an electricity cost of $0.001 per hour.

For how many hours of use it is cheaper to use an energy saver light bulb than a regular light bulb?

a) Write an inequality for this problem.

b) Solve the inequality. Explain the solution in words.

4.D. Linear inequalities with two variables

In this case we are working with x and y. At the same time the solutions are all the points that are situated in a certain area of a plane defined by the cartesian coordinates of the points.

EXAMPLE

Solve the inequality

$y < 2x + 1$

We have to find all the points in the plane that satisfy this condition.

Step 1:

Represent the straight line $y = 2x + 1$

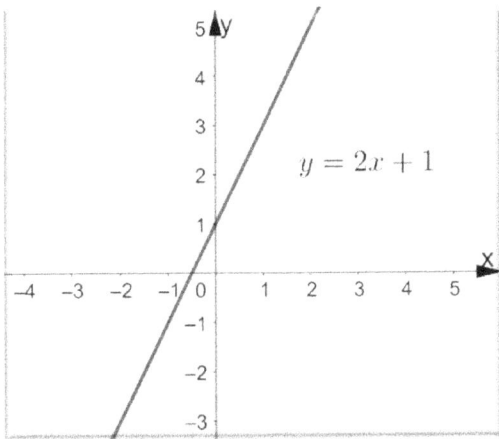

Step 2:

Check whether point (0,0) satisfies the inequality:

$y < 2x + 1$

We substitute zero for x and y respectively. We have:

$0 < 2(0) + 1$

$0 < 1$

Because this is true, it means that point (0,0) is part of the solution.

The fact that y is less than 2x+1 means that the points situated on the line $y = 2x + 1$ are not part of the solution.

We represent the final solution as the area situated below the line $y = 2x + 1$

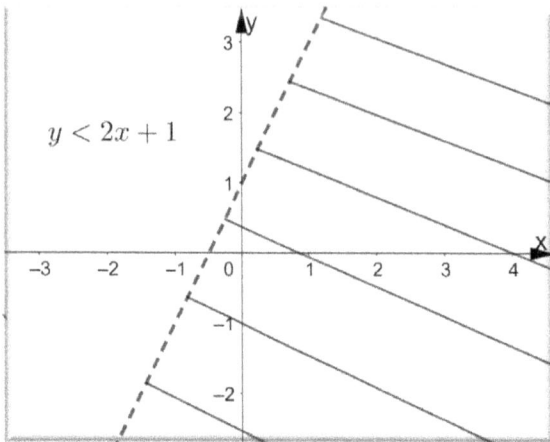

$y < 2x + 1$

The line is a dashed line because the points on this line are not part of the solution. All the other points beneath the line $y = 2x + 1$ are part of the solution.

EXAMPLE

Solve the inequality

$y \geq -3x + 2$

We have to find all the points in the plane that satisfy this condition.

Step 1:

Represent the straight line $y = -3x + 2$

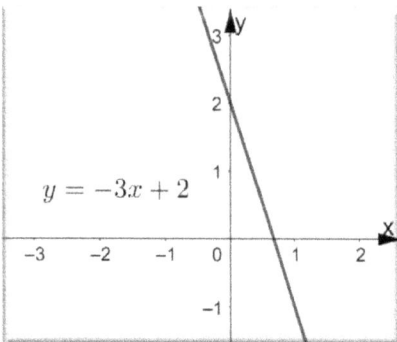

$y = -3x + 2$

Step 2:

Check whether point (0,0) satisfies the inequality:

$y \geq -3x + 2$

We substitute zero for x and y respectively.

We have:

$0 \geq -3(0) + 2$

$0 \geq 2$

Because this is not true, it means that point (0,0) is not part of the solution.

The fact that y is greater and equal than -3x+2 means that the points situated on the line

$y = -3x + 2$ are part of the solution. The line will be a full line not dashed.

We represent the final solution as the area situated below the line $y \geq -3x + 2$

All the other points above the line $y = -3x + 2$ will be part of the solution.

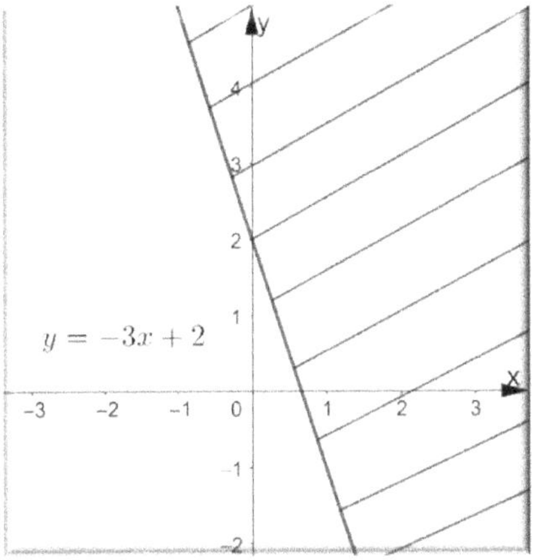

$y = -3x + 2$

> **Remember:**
> < or > means that the solution region does NOT include the points on the line that is the boundary
> $\leq$ or $\geq$ means that the solution region INCLUDES the points on the line that is the boundary

PRACTICE

1) Graph the solution

$y > 3x - 1$

2) Which point(s) is/are in the solution region of the inequality $3x - 6y \leq 4$

a) (0,0)   b) (4,2)   c) (-2,5)   d) (3,-6)

3) Sketch the graphs of the following inequalities.

a) $y \geq 0$ b) $x > 2 \ and \ y > 1$

4) Graph the solution of two variable inequality

$4x + 3y \geq -9$

5) In the graph below, the equation of the boundary line is: $x - 3y = 6$
Determine the inequality represented by the graph.

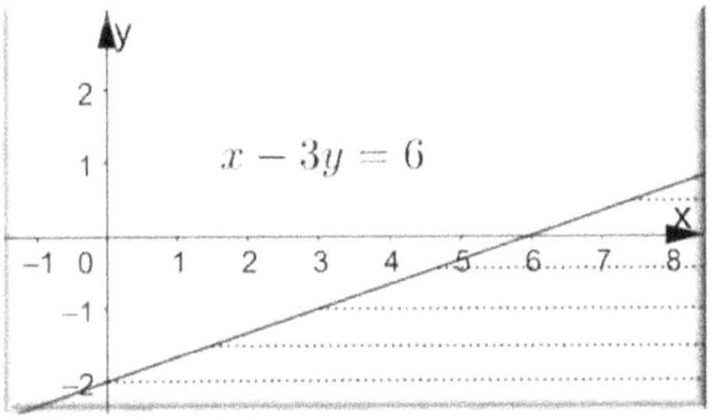

$x - 3y = 6$

6) The graph below shows the solution to the inequality

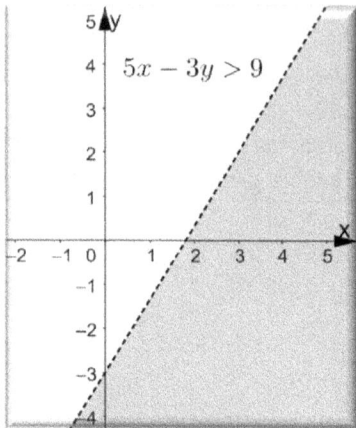

a) Explain why the boundary line is a broken line?

b) Why the solution is beneath the line and not above?

7) Without showing any work, sketch the graph of the following inequality:

$y + 2 < 0$

8) Show the solution region to the inequality: $3x + 4y > 8$

9) The equation of the boundary is given. Determine the inequality which is represented by the solution region.

$3x - 5y + 15 = 0$

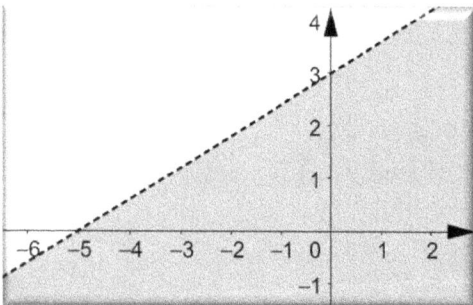

10) The point(s) which is NOT in the solution region of the inequality $5x - 2y > 5$ are:

a) (1,3)   b) (3,2)

c) (2,-3)   d(0,1)

STEP BY STEP SOLUTIONS

1.B. Ordered pairs - Introduction

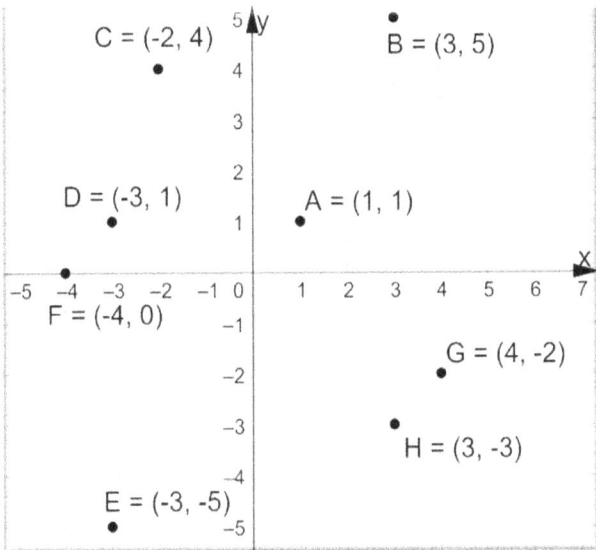

In what quadrants are situated the points below?

A (1,1)   Q1   B (3,5) Q1
C (-2,4)   Q2   D (-3,1) Q2
E (-3,-5) Q3   F (-4,0) Q2
G (4,-2)   Q4   H (3,-3) Q4

2.A. Representing patterns in linear relations

1) Analyze the pattern that figures below have. Find how many houses figure 5 will have.

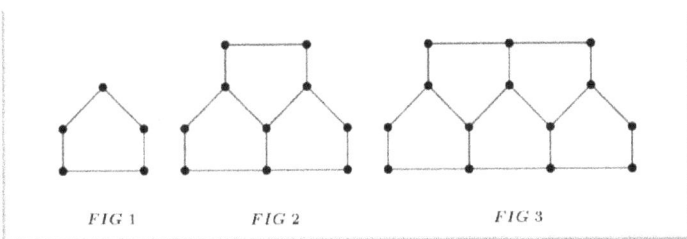

FIG 1   FIG 2   FIG 3

Figure	# Houses
1	1
2	3
3	5
5	9

2) Analyze the pattern that figures below have. Find how many sticks figure 6 will have.

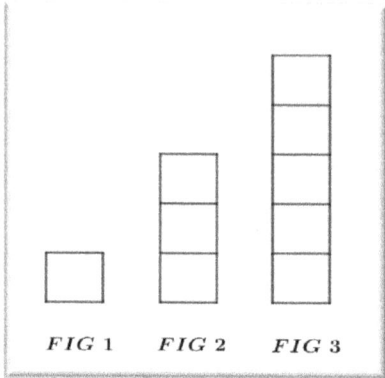

Figure	# Squares
1	1
2	3
3	5
4	7
6	11

3) Analyze the pattern that figures below have. Find how many squares figure 5 will have.

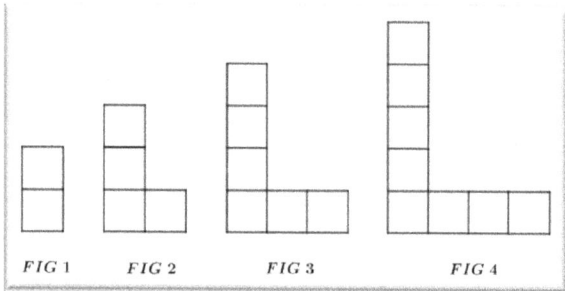

Figure	# Squares
1	2
2	4
3	6
4	8
5	10

4) Analyze the pattern that figures below have. Find how many squares figure 6 will have.

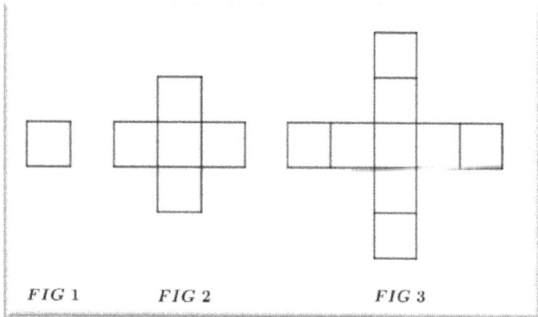

Figure	# Squares
1	1
2	5
3	9
4	13
6	21

5) Analyze the pattern that figures below have. Find how many triangles figure 5 will have.

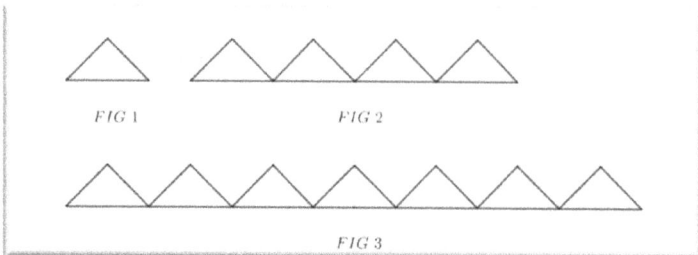

Figure	# Triangles
1	1
2	4
3	7
4	10
5	13

6) Analyze the pattern that the table below has. Find how many figurines the 5th child will have.

Child	# Figurines
1	3
2	7
3	11
4	15
5	19

7) For each piece she needs 6 carboard pieces.

She needs to build 6 more.

$\frac{6\ pieces}{cube} \times 6\ cubes = 36\ pieces$

She has only 31 pieces.

She can finish only another 5 cubes with a total of nine to be finished not ten cubes.

Cube	# Carboard pieces
1	6
2	12
3	18
4	24

8) Analyze the pattern that the figures below have. Find how many squares the 5th figure will have.

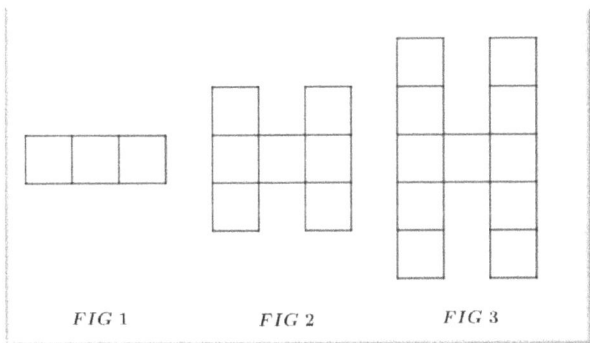

FIG 1   FIG 2   FIG 3

Figure	# Squares
1	3
2	7
3	11
4	15
5	19

9) Make a table of values for the first four terms. Start at 4. To get the next term, triple the number and subtract 1.

X	Y=3x-1
4	Y=3(4)-1=11
5	Y=3(5)-1=14
6	Y=3(6)-1=17
7	Y=3(7)-1=20

10)

$Cost = 40 + 5 \times 12 = 40 + 60 = \$100$

2.B. Understanding tables of values of linear relationships

1) In the relation $B = 4q$, determine B when q is:

a) $B = 4q = 4 \times 2 = 8$ b) $B = 4q = 4 \times 13 = 52$

c) $B = 4q = 4 \times (-5) = -20$ d) $B = 4q = 4 \times (x + 2) = 4x + 8$

2) In the relation $M = 3k - 4$ determine M when k is:

a) $M = 3k - 4 = 3(2) - 4 = 6 - 4 = 2$

b) $M = 3k - 4 = 3(17) - 4 = 51 - 4 = 47$

c) $M = 3k - 4 = 3(40) - 4 = 120 - 4 = 116$

d) $M = 3k - 4 = 3(y + 3) - 4 = 3y + 9 - 4 = 3y + 5$

3)

a) 5-2=3   8-5=3   11-8=3   Common difference is: 3

b) , $2\sqrt{3} - \sqrt{3} = \sqrt{3}$ $3\sqrt{3} - 2\sqrt{3} = \sqrt{3}$ $4\sqrt{3} - 3\sqrt{3} = \sqrt{3}$

Common difference is $\sqrt{3}$

4) Determine the common difference in the pattern below.

$\sqrt{3} + \sqrt{5}, \quad \sqrt{3}, \quad \sqrt{3} - \sqrt{5} \dots \dots \dots$

$\sqrt{3} - (\sqrt{3} + \sqrt{5}) = \sqrt{3} - \sqrt{3} - \sqrt{5} = -\sqrt{5}$

$\sqrt{3} - \sqrt{5} - \sqrt{3} = -\sqrt{5}$

Common difference is $-\sqrt{5}$

5) Analyze the table below and write a relation between x and y.

X	1	2	3	4	5
Y	6	9	12	15	18

Step 1

Check the difference between two consecutive values in the y column.

In this case that difference is 3

Step 2

Form the equation $y = 3 \times x$

Step 3

Check if for $x = 1$, the value for y at step 2 equals the value we should get in column y.

For $x = 1$, $y = 3 \times (1) = 3$

Step 4

We add or subtract any value from $A \times x$ in such a way that we obtain the value of y that is beside the value of x in the x column.

The value of y that corresponds to x=1 is 6 not 3.

We have to add 3 units to $y = 3 \times x$ in order to get to 6.

The relation between y and x in this case will be:

$y = 3 \times x + 3$

Step 5

Check for another value of x if we obtain the correct corresponding value of y.

Let us take $x = 3$

Then we have: $y = 3 \times (3) + 3 = 9 + 3 = 12$ which equals the value of y for $x = 3$.

The equation that connects x and y is indeed $y = 3x + 3$

6) Analyze the table below and write a relation between x and y.

X	1	2	3	4	5
Y	15	10	5	0	-5

Step 1

Check the difference between two consecutive values in the y column.

In this case that difference is -5

Step 2

Form the equation $y = -5 \times x$

Step 3

Check if for $x = 1$, the value for y at step 2 equals the value we should get in column y.

For $x = 1$, $y = -5 \times (1) = -5$

Step 4

We add or subtract any value from $A \times x$ in such a way that we obtain the value of y that is beside the value of x in the x column.

The value of y that corresponds to x=1 is 15 not -5.

We have to add 20 units to $y = -5 \times x$ in order to get 15.

The relation between y and x in this case will be:

$y = -5 \times x + 20$

Step 5

Check for another value of x if we obtain the correct corresponding value of y.

Let us take $x = 3$

Then we have: $y = -5 \times (3) + 20 = -15 + 20 = 5$ which equals the value of y for $x = 3$.

The equation that connects x and y is indeed $y = -5x + 20$

7) Analyze the table below and write a relation between x and y.

X	1	2	3	4	5
Y	-5	-1	3	7	11

Step 1

Check the difference between two consecutive values in the y column.

In this case that difference is 4

Step 2

Form the equation $y = 4 \times x$

Step 3

Check if for $x = 1$, the value for y at step 2 equals the value we should get in column y.

For $x = 1$, $y = 4 \times (1) = 4$

Step 4

We add or subtract any value from $A \times x$ in such a way that we obtain the value of y that is beside the value of x in the x column.

The value of y that corresponds to x=1 is 4 not -5.

We have to subtract 9 units to $y = 4 \times x$ in order to get -5.

The relation between y and x in this case will be:

$y = 4 \times x - 9$

Step 5

Check for another value of x if we obtain the correct corresponding value of y.

Let us take $x = 3$

Then we have: $y = 4 \times (3) - 9 = 12 - 9 = 3$ which equals the value of y for $x = 3$.

The equation that connects x and y is indeed $y = 4x - 9$

8) Determine the 20th term of the billow linear pattern

5,9,13,17,21

We create the table:

X (term #)	1	2	3	4	5
Y	5	9	13	17	21

Step 1

Check the difference between two consecutive values in the y column.

In this case that difference is 4

Step 2

Form the equation $y = 4 \times x$

Step 3

Check if for $x = 1$, the value for y at step 2 equals the value we should get in column y.

For $x = 1$, $y = 4 \times (1) = 4$

Step 4

We add or subtract any value from $A \times x$ in such a way that we obtain the value of y that is beside the value of x in the x column.

The value of y that corresponds to x=1 is 4 not 5.

We have to add 1 unit to $y = 4 \times x$ in order to get 5.

The relation between y and x in this case will be:

$y = 4 \times x + 1$

Step 5

Check for another value of x if we obtain the correct corresponding value of y.

Let us take $x = 3$

Then we have: $y = 4 \times (3) + 1 = 12 + 1 = 13$ which equals the value of y for $x = 3$.

The equation that connects x and y is indeed $y = 4x + 1$

The 50th term will be for x=50
$y(50) = 4(50) + 1 = 200 + 1 = 201$

9) The total cost to publish a book, for a publishing company, is a fixed price (100) plus a cost for each book that the company will print. Create a general relation between the number printed books and the cost of printing.

X (# of Books)	0	100	200	300	400
Cost	100	300	500	700	900

If we consider the situation for one book, two books and so on we will have the table:

X (# of Books)	0	1	2	3	4
Cost	100	3	5	7	9

Step 1
Check the difference between two consecutive values in the y column.
In this case that difference is 2
Step 2
Form the equation $y = 2 \times x$
Step 3
Check if for $x = 1$, the value for y at step 2 equals the value we should get in column y.
For $x = 1, y = 2 \times (1) = 2$
Step 4
We add or subtract any value from $A \times x$ in such a way that we obtain the value of y that is beside the value of x in the x column.
The value of y that corresponds to x=1 is 2 not 3.
We have to add 1 unit to $y = 2 \times x$ in order to get 3.
The relation between y and x in this case will be:
$y = 2 \times x + 1$
Step 5
Check for another value of x if we obtain the correct corresponding value of y.
Let us take $x = 3$
Then we have: $y = 2 \times (3) + 1 = 6 + 1 = 7$ which equals the value of y for $x = 3$.

The relation that connects the number of books printed and cost is indeed $y = 2x + 1$ plus the initial cost of $100

10) Using the same data from problem 9, determine the cost of printing 543 books.

The relation that connects the number of books printed and cost is indeed $y = 2x + 1$ plus the initial cost of $100.

If the number of books is 345, the total cost will e calculated by:

$$y = 2x + 1 + 100 = 2(345) + 101 = 690 + 101 = \$791$$

2.C. Understanding graphs of linear relationships

1) Represent the points from the table below and see if they are part of a straight line.

Point	X	Y
A	1	2
B	2	4
C	3	6
D	4	8

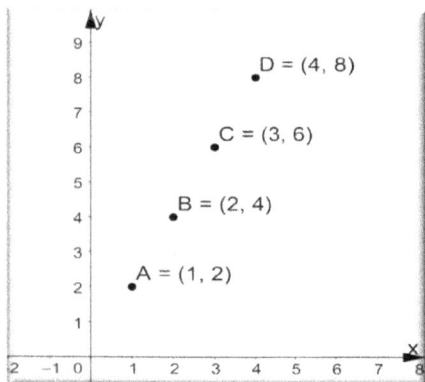

2) Represent the points from the table below and see if they are part of a straight line.

Point	X	Y
A	1	-4
B	2	0
C	3	4
D	4	8

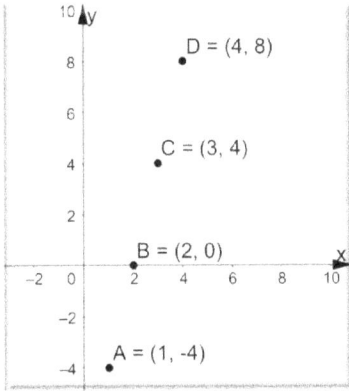

3) Represent the points from the table below and see if they are part of a straight line.

Point	X	Y
A	1	1
B	2	2
C	3	3

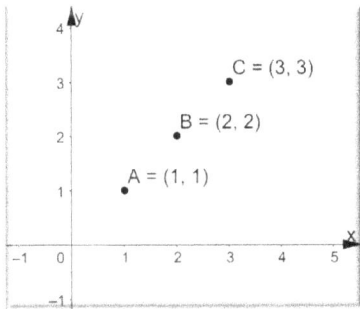

4) Represent the points from the table below and see if they are part of a straight line.

Point	X	Y
A	1	-3
B	2	-1
C	3	1
D	4	3

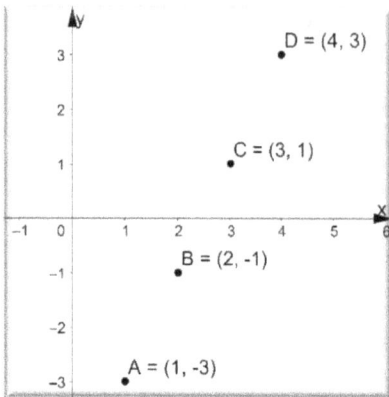

5) Represent the points from the table below and see if they are part of a straight line.

Point	X	Y
A	1	2
B	2	3
C	3	5
D	4	8

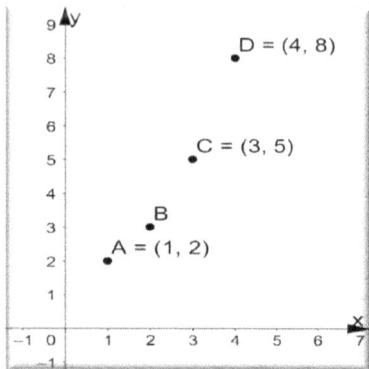

6) Represent the points from the table below and see if they are part of a straight line.

Point	X	Y
A	1	2
B	2	4
C	4	5
D	5	8

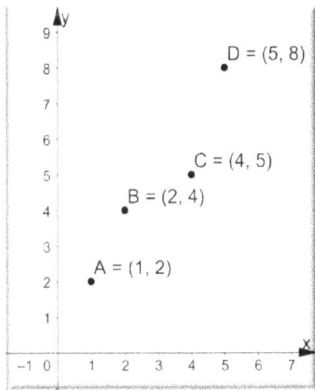

7) Represent the points from the table below and see if they are part of a straight line.

Point	X	Y
A	1	2
B	2	-1
C	3	6
D	4	-3

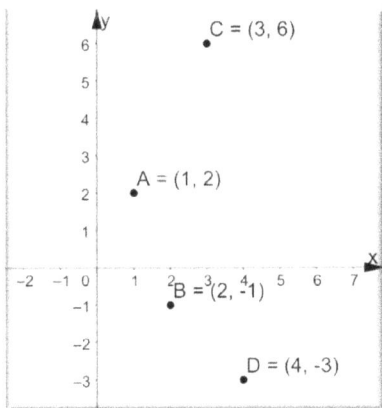

8) Represent the points from the table below and see if they are part of a straight line.

Point	X	Y
A	1	1
B	2	3
C	3	5

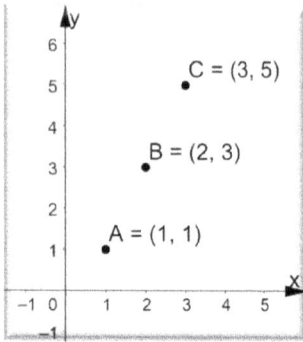

9) Represent the points from the table below and see if they are part of a straight line.

Point	X	Y
A	-1	2
B	2	4
C	-3	6

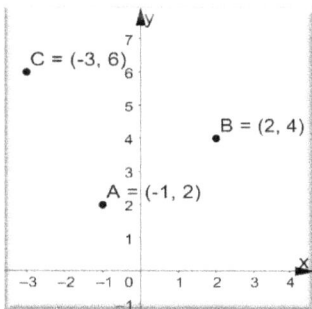

10) Represent the points from the table below and see if they are part of a straight line.

Point	X	Y
A	1	2
B	3	-3
C	5	6
D	4	7

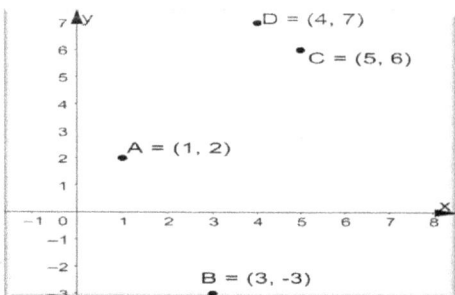

.D. Distance between points

. Horizontal distance

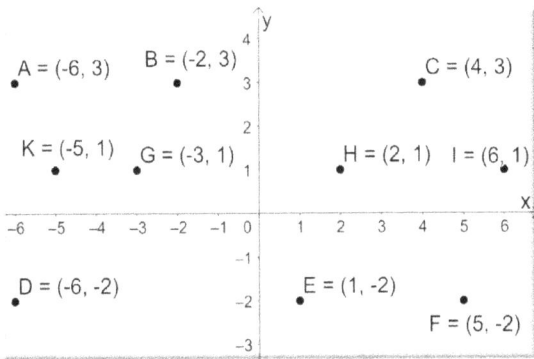

1) Correct

The distance between point A and point C is 10

$|4 - (-6)| = |10| = 10$

2) Incorrect

 The distance between point A and point B is 5

$|-2 - (-6)| = |4| = 4$

3) Correct

The distance between point D and point E is 7

$|1 - (-6)| = |7| = 7$

) Incorrect

he distance between point E and point F is 7

$; - (1.| = |4| = 4$

) Correct

he distance between point K and point G is 2

$-3 - (-5)| = |2| = 2$

) Correct

he distance between point K and point H is 7

$2 - (-5)| = |7| = 7$

) Incorrect

he distance between point K and point I is 9

$; - (-5)| = |11| = 11$

) Correct

he distance between point G and point H is 5

$2 - (-3)| = |5| = 5$

) Incorrect

he distance between point G and point I is 8

$; - (-3)| = |9| = 9$

10) Correct

The distance between point H and point I is 4

$|6 - 2| = |4| = 4$

2.D Distance between points
b. Vertical distance

1) Incorrect

The distance between point A and point K is 10

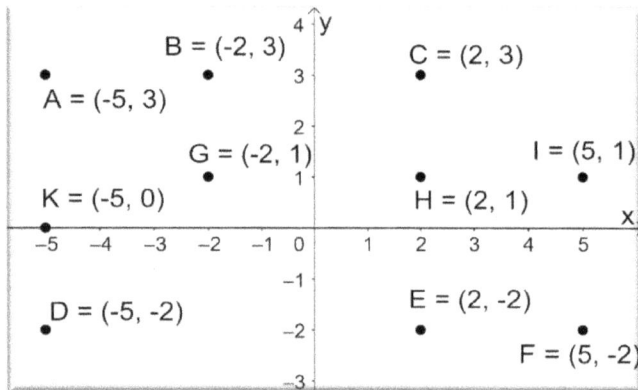

$|0 - 3| = |-3| = 3$

2) Correct

The distance between point A and point D is 5

$|-2 - 3| = |-5| = 5$

3) Incorrect

The distance between point B and point G is 3

$|1 - 3| = |-2| = 2$

4) Incorrect

The distance between point C and point H is 7

$|1 - 3| = |-2| = 2$

5) Correct

The distance between point C and point E is 5

$|-2 - 3| = |-5| = 5$

6) Incorrect

The distance between point K and point D is 7

$|-2 - (0)| = |-2| = 2$

7) Correct

The distance between point H and point E is 3

$|-2 - 1| = |-3| = 3$

8) Incorrect

The distance between point I and point F is 7

$|-2 - (-6)| = |4| = 4$

9) Incorrect

The distance between point A and point K is 9

$|0 - 3| = |-3| = 3$

10) Incorrect

The distance between point I and point H is 4

$|5 - 2| = |3| = 3$ (*Horizontal distance*)

2.D Distance between points

c. Non-horizontal and Non-vertical distance

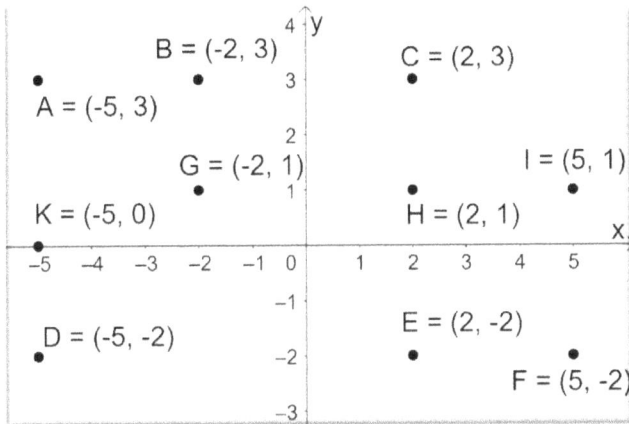

1) Correct

The distance between point B and point H is

$2\sqrt{5} = 4.47$

BH=$\sqrt{(x_2 - x_1)^2 + (y_2 - y_1)^2} =$

$\sqrt{[(2 - (-2)]^2 + (1 - 3)^2} = \sqrt{16 + 4} = \sqrt{20} =$

$\sqrt{4 \times 5} = 2\sqrt{5}$

2) Incorrect

The distance between point K and point C is

7.61

GC=$\sqrt{(x_2 - x_1)^2 + (y_2 - y_1)^2} = \sqrt{(2 - (-5))^2 + (3 - 0)^2} = \sqrt{49 + 9} = \sqrt{58} = 7.61$

3) Correct

The distance between point A and point G is 3.6

AG=$\sqrt{(x_2 - x_1)^2 + (y_2 - y_1)^2} = \sqrt{(-2 - (-5))^2 + (1 - 3)^2} = \sqrt{9 + 4} = \sqrt{13} = 3.6$

4) Correct

The distance between point A and point H is

AH=$\sqrt{(x_2 - x_1)^2 + (y_2 - y_1)^2} = \sqrt{(2 - (-5))^2 + (1 - 3)^2} = \sqrt{49 + 4} = \sqrt{53} = 7.28$

5) Incorrect

The distance between point A and point I is 10.19

AI=$\sqrt{(x_2 - x_1)^2 + (y_2 - y_1)^2} = \sqrt{(5 - (-5))^2 + (1 - 3)^2} = \sqrt{100 + 4} = \sqrt{104} = 10.19$

6) Correct

The distance between point A and point F is 11.18

$$AF = \sqrt{(x_2 - x_1)^2 + (y_2 - y_1)^2} = \sqrt{(5 - (-5))^2 + (-2 - 3)^2} = \sqrt{100 + 25} = \sqrt{125} = 11.18$$

7) Correct

The distance between point A and point E is 8.6

$$AE = \sqrt{(x_2 - x_1)^2 + (y_2 - y_1)^2} = \sqrt{(2 - (-5))^2 + (-2 - 3)^2} = \sqrt{49 + 25} = \sqrt{74} = 8.6$$

8) Incorrect

The distance between point B and point D is 5.83

$$BD = \sqrt{(x_2 - x_1)^2 + (y_2 - y_1)^2} = \sqrt{(-5 - (-2))^2 + (-2 - 3)^2} = \sqrt{9 + 25} = \sqrt{34} = 5.83$$

9) Correct

The distance between point B and point E is 6.4

$$BE = \sqrt{(x_2 - x_1)^2 + (y_2 - y_1)^2} = \sqrt{(2 - (-2))^2 + (-2 - 3)^2} = \sqrt{16 + 25} = \sqrt{41} = 6.4$$

10) Incorrect

The distance between point B and point F is 8.6

$$BF = \sqrt{(x_2 - x_1)^2 + (y_2 - y_1)^2} = \sqrt{(5 - (-2))^2 + (-2 - 3)^2} = \sqrt{49 + 25} = \sqrt{74} = 8.6$$

2.D Distance between points
d. Midpoint coordinates

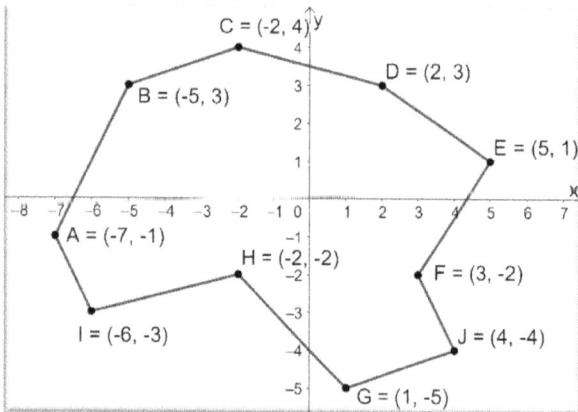

1) Correct

The mid-point coordinates of segment AB are:

$$x = \frac{x_2 + x_1}{2} = \frac{-5 + (-7)}{2} = \frac{-12}{2} = -6$$

$$y = \frac{y_2 + y_1}{2} = \frac{3 + (-1.)}{2} = \frac{2}{2} = 1$$

2) Incorrect

The mid-point coordinates of segment BC are:

$$x = \frac{x_2 + x_1}{2} = \frac{-2 + (-5)}{2} = \frac{-7}{2} = -3.5$$

$$y = \frac{y_2 + y_1}{2} = \frac{4 + 3}{2} = \frac{7}{2} = 3.5$$

3) Correct

The mid-point coordinates of segment CD are:

$$x = \frac{x_2 + x_1}{2} = \frac{2 + (-2)}{2} = \frac{0}{2} = 0$$

$$y = \frac{y_2 + y_1}{2} = \frac{4 + 3}{2} = \frac{7}{2} = 3.5$$

4) Incorrect

The mid-point coordinates of segment DE are:

$$x = \frac{x_2 + x_1}{2} = \frac{2+5}{2} = \frac{7}{2} = 3.5$$

$$y = \frac{y_2 + y_1}{2} = \frac{1+3}{2} = \frac{4}{2} = 2$$

5) Incorrect

The mid-point coordinates of segment EF are:

$$x = \frac{x_2 + x_1}{2} = \frac{3+5}{2} = \frac{8}{2} = 4$$

$$y = \frac{y_2 + y_1}{2} = \frac{-2+1}{2} = \frac{-1}{2} = -0.5$$

6) Correct

The mid-point coordinates of segment FJ are:

$$x = \frac{x_2 + x_1}{2} = \frac{3+4}{2} = \frac{7}{2} = 3.5$$

$$y = \frac{y_2 + y_1}{2} = \frac{-2-4}{2} = \frac{-6}{2} = -3$$

7) Correct

The mid-point coordinates of segment JG are:

$$x = \frac{x_2 + x_1}{2} = \frac{1+4}{2} = \frac{5}{2} = 2.5$$

$$y = \frac{y_2 + y_1}{2} = \frac{-5+(-4)}{2} = \frac{-9}{2} = -4.5$$

8) Incorrect

The mid-point coordinates of segment GH are:

$$x = \frac{x_2 + x_1}{2} = \frac{-2+1}{2} = \frac{-1}{2} = -0.5$$

$$y = \frac{y_2 + y_1}{2} = \frac{-2+(-5)}{2} = \frac{-7}{2} = -3.5$$

9) Correct

The mid-point coordinates of segment HI are:

$$x = \frac{x_2 + x_1}{2} = \frac{-6+(-2)}{2} = \frac{-8}{2} = -4$$

$$y = \frac{y_2 + y_1}{2} = \frac{-3+(-2)}{2} = \frac{-5}{2} = -2.5$$

10) Incorrect

The mid-point coordinates of segment IA are:

$$x = \frac{x_2 + x_1}{2} = \frac{-7+(-6)}{2} = \frac{-13}{2} = -6.5 \quad y = \frac{y_2 + y_1}{2} = \frac{-1+(-3)}{2} = \frac{-4}{2} = -2$$

2.E Slope of a line

1) Correct

The slope of segment AB is:

$$m = \frac{y_2 - y_1}{x_2 - x_1} = \frac{3-(-1.}{-5-(-7)} = \frac{4}{2} = 2$$

2) Incorrect

The slope of segment BC is

$$m = \frac{y_2 - y_1}{x_2 - x_1} = \frac{4-3}{-2-(-5)} = \frac{1}{3}$$

3) Correct

The slope of segment CD is

$$m = \frac{y_2 - y_1}{x_2 - x_1} = \frac{3-4}{2-(-2)} = \frac{-1}{4}$$

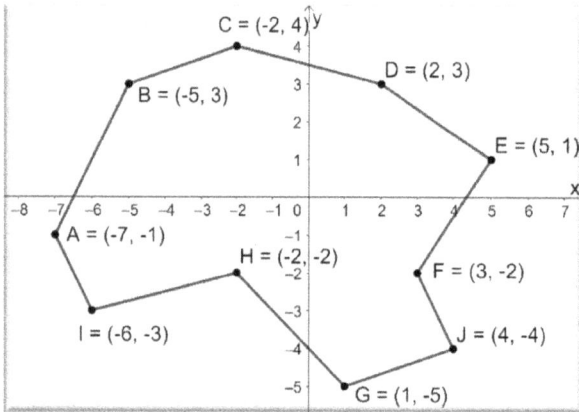

4) Incorrect

The slope of segment DE is

$$m = \frac{y_2 - y_1}{x_2 - x_1} = \frac{1-3}{5-2} = \frac{-2}{3}$$

5) Incorrect

The slope of segment EF is

$$m = \frac{y_2 - y_1}{x_2 - x_1} = \frac{-2-1}{3-5} = \frac{-3}{-2} = 1.5$$

6) Correct

The slope of segment FJ is:

$$m = \frac{y_2 - y_1}{x_2 - x_1} = \frac{-4-(-2)}{4-3} = \frac{-2}{1} = -2$$

7) Correct

The slope of segment JG is

$$m = \frac{y_2 - y_1}{x_2 - x_1} = \frac{-5-(-4)}{1 \quad 4} = \frac{-1}{-3} = \frac{1}{3}$$

8) Incorrect

The slope of segment GH is

$$m = \frac{y_2 - y_1}{x_2 - x_1} = \frac{-2-(-5)}{-2-1} = \frac{3}{-3} = -1$$

9) Correct

The slope of segment HI is

$$m = \frac{y_2 - y_1}{x_2 - x_1} = \frac{-3-(-2)}{-6-(-2)} = \frac{-1}{-4} = \frac{1}{4}$$

10) Incorrect

The slope of segment IA is

$$m = \frac{y_2 - y_1}{x_2 - x_1} = \frac{-1-(-3)}{-7-(-6)} = \frac{2}{-1} = -2$$

3.B. Solving Linear Equations

a. Solve one-step linear equations:

1) Solve ad check the solution.

a) $x + 3 = 25$

b) $x + \frac{1}{2} = 3\frac{1}{3}$

Solution

a) $x + 3 - 3 = 25 - 3$ We subtract 3 on each side

$x + 0 = 22$

CHECK

$22 + 3 = 25$

$25 = 25$

b) $x + \frac{1}{2} = 3\frac{1}{3}$ We subtract $\frac{1}{2}$ from both sides of the equation

$x + \frac{1}{2} - \frac{1}{2} = 3\frac{1}{3} - \frac{1}{2}$

$x + 0 = 3\frac{1}{3} - \frac{1}{2}$ We transform the mixed fraction into an improper fraction

$x = \frac{10}{3} - \frac{1}{2}$ We find the common denominator as 6

$x = \frac{10 \times 2}{3 \times 2} - \frac{1 \times 3}{2 \times 3}$

$x = \frac{20}{6} - \frac{3}{6}$

$x = \frac{20 - 3}{6} = \frac{17}{6} = 2\frac{5}{6}$

CHECK

$\frac{17}{6} + \frac{1}{2} = 3\frac{1}{3}$ We find that the common denominator is 6

$\frac{17}{6} + \frac{1 \times 3}{2 \times 3} = 3\frac{1}{3}$

$\frac{17}{6} + \frac{3}{6} = 3\frac{1}{3}$

$\frac{17 + 3}{6} = 3\frac{1}{3}$

$\frac{20}{6} = 3\frac{1}{3}$

$3\frac{2}{6} = 3\frac{1}{3}$ We divide fraction $\frac{2}{6}$ with 2 at the denominator and numerator

$3\frac{1}{3} = 3\frac{1}{3}$

2) Write each sentence as an equation

a) x decreased by 3 is 7   b) The sum of 4 and x is 9

$x - 3 = 7$ $x + 4 = 9$

3) Check if the given value of x is a solution for the equation.

a) $x + 3 = 7$ for $x = 5$ b) $-x - 2 = 7$ for $x = 4$

c) $x + 5 = 7$ for $x = 2$ d) $x + 3.4 = 5.8$ for $x = 4.3$

a) $5 + 3 = 7$

$8 \neq 7$ $x = 5$ is not a solution

b) $-4 - 2 = 7$

$-6 \neq 7$ $x = 4$ is not a solution

c) $2 + 5 = 7$

$7 = 7$ $x = 2$ is the solution

d) $4.3 + 3.4 = 5.8$

$7.7 \neq 5.8$ $x = 4.3$ is not a solution

4) What is the step needed to isolate the variable?

a) $x + 4 = 5$

b) $x - 3 = 15$

c) $4 = 7 + x$

d) $-4 = 6 + x$

a) subtract 4 on each side

b) add 3 on each side

c) subtract 7 on each side

d) subtract 6 on each side

5) Explain the error.

$$x - \frac{2}{7} = \frac{5}{7}$$

$$x = \frac{3}{7}$$

Error: we have to add $\frac{2}{7}$ on both sides. So, we will have $x = \frac{2}{7} + \frac{5}{7} = \frac{7}{7} = 1$

6) Write an equation. Solve ad check.

Five more than n is negative seen

Equation

$5 + n = -7$ we subtract 5 on each side of the equation.

$5 - 5 + n = -7 - 5$

$0 + n = -12$

$n = -12$

CHECK

$5 - 12 = -7$

$-7 = -7$ $n = -12$ is the solution

7) The length and the width of a rectangle add to 35 cm. if the width is 10 cm, what is the length?

$L + W = 35$

$W = 10$

So,

$L + 10 = 35$ we subtract 10 on both sides.

$L + 10 - 10 = 35 - 10$

$L + 0 = 25$

The Length is 25 cm

8) A book price was reduced by $3.25 and then sold for $7.27. What was the original price of the book?

P is the original price.

$P - 3.25 = 7.27$ We add 3.25 o each side.

$P - 3.25 + 3.25 = 7.27 + 3.25$

$P - 0 = 10.52$ The original price was $10.52

9) A right triangle has one of the angles (A) equal with 25 degrees. How big is the other angle beside the right one?

The sum of the interior angles in a triangle is 180^0. The right angle is 90^0

The equation is:

$A + 90^0 + 25^0 = 180^0$

$A + 115^0 = 180^0$ We subtract 115^0 on both sides of the equation.

$A + 115^0 - 115^0 = 180^0 - 115^0$

$A + 0 = 65^0$

$A = 65^0$

10) The 30cm Length of a rectangle is bigger that the Width by 12cm. How much is the Perimeter?

So,

$30 - 12 = W$

$W = 18 cm$

$Perimeter = 2(30) + 2(18) = 60 + 36 = 96 cm$

3.B. Solving Linear Equations

b. Solving two-step linear equations with addition and subtraction

1) Solve and check

$2x - 4 = x + 2$

$2x - 4 = x + 2$ We add 4 on each side

$2x - 4 + 4 = x + 2 + 4$

$2x - 0 = x + 6$ We subtract x on each side

$2x - x = x - x + 6$

$x = 0 + 6$

$x = 6$

CHECK

$2(6) - 4 = 6 + 2$

$12 - 4 = 8$

$8 = 8$

$x = 6$ is the solution of the equation $2x - 4 = x + 2$

2) Solve and check.

$5x + \frac{1}{3} = 4x + 1\frac{1}{2}$

$5x + \frac{1}{3} = 4x + 1\frac{1}{2}$	We subtract $\frac{1}{3}$ on each side
$5x + \frac{1}{3} - \frac{1}{3} = 4x + 1\frac{1}{2} - \frac{1}{3}$	
$5x + 0 = 4x + 1\frac{1}{2} - \frac{1}{3}$	We subtract $4x$ on each side
$5x - 4x = 4x - 4x + 1\frac{1}{2} - \frac{1}{3}$	
$x = 0 + 1\frac{1}{2} - \frac{1}{3}$	We transform $1\frac{1}{2}$ into an improper fraction
$x = \frac{3}{2} - \frac{1}{3}$	We find common denominator being 6
$x = \frac{3\times3}{2\times3} - \frac{1\times2}{3\times2}$	
$x = \frac{9}{6} - \frac{2}{6} = \frac{9-2}{6} = \frac{7}{6} = 1\frac{1}{6}$	
$x = 1\frac{1}{6}$	

CHECK

$5(\frac{7}{6}) + \frac{1}{3} = 4(\frac{7}{6}) + 1\frac{1}{2}$	We transform $1\frac{1}{2}$ into an improper fraction
$5(\frac{7}{6}) + \frac{1}{3} = 4(\frac{7}{6}) + \frac{3}{2}$	We multiply 5 and 4 with $\frac{7}{6}$
$\frac{35}{6} + \frac{1}{3} - \frac{28}{6} + \frac{3}{2}$	We multiply each term with 6 The common denominator
$\frac{35\times6}{6} + \frac{1\times6}{3} = \frac{28\times6}{6} + \frac{3\times6}{2}$	We simplify wherever possible.
$35 + 2 = 28 + 9$	
$37 = 37$	

$x = 1\frac{1}{6}$ is the solution of the equation $5x + \frac{1}{3} = 4x + 1\frac{1}{2}$

3) Solve ad check

$3.25x - 5 = 2.25x - 6$

$3.25x - 5 = 2.25x - 6$	We add 5 on each side
$3.25x - 5 + 5 = 2.25x - 6 + 5$	
$3.25x + 0 = 2.25x - 1$	We subtract $2.25\,x$ on each side
$3.25x - 2.25x = 2.25x - 2.25x - 1$	
$x = -1$	

CHECK

$3.25(-1) - 5 = 2.25(-1) - 6$
$3.25(-1) - 5 = 2.25(-1) - 6$
$-3.25 - 5 = -2.25 - 6$
$-8.25 = -8.25$
$x = -1$ is the solution of the equation $3.25x - 5 = 2.25x - 6$

4) Check if these given value of x are the solution for the equation below.
$3x - 5 = 5x + 3$ $\qquad$ $x = 2, -3, -4$

For $x = 2$
$3(2) - 5 = 5(2) + 3$
$6 - 5 = 10 + 3$
$1 \neq 13$ $\qquad$ $x = 2$ is not a solution of the equation $3x - 5 = 5x + 3$

For $x = -3$
$3(-3) - 5 = 5(-3) + 3$
$-9 - 5 = -15 + 3$
$-14 \neq -12$
$x = -3$ is not a solution for the equation $3x - 5 = 5x + 3$

For $x = -4$
$3(-4) - 5 = 5(-4) + 3$
$-12 - 5 = -20 + 3$
$-17 = -17$
$x = -4$ is a solution for the equation $3x - 5 = 5x + 3$

5) Solve ad check
$2x + 4x - 4 = 5x - 3$ $\qquad$ We add the terms that have x (like 2apples+4apples)

$6x - 4 = 5x - 3$ $\qquad$ We add 4 on both sides of the equation
$6x - 4 + 4 = 5x - 3 + 4$
$6x - 0 = 5x + 1$ $\qquad$ We subtract $5x$ on both sides of the equation
$6x - 5x = 5x - 5x + 1$
$x = 1$

CHECK

$2(1) + 4(1) - 4 = 5(1) - 3$
$2 + 4 - 4 = 5 - 3$
$2 = 2$
$x = 1$ is the solution of the equation $2x + 4x - 4 = 5x - 3$

6) Write an equation. Solve and check.

Four times a number increased by 3 is 3 times a number decreased by 1

$4x + 3 = 3x - 1$ We subtract 3 on both sides.

$4x + 3 - 3 = 3x - 1 - 3$

$4x = 3x - 4$ We subtract 3x on each side

$4x - 3x = 3x - 3x - 4$

$x = -4$

CHECK

$4(-4) + 3 = 3(-4) - 1$

$-16 + 3 = -12 - 1$

$-13 = -13$

$x = -4$ is the solution of the equation $4x + 3 = 3x - 1$

7) Write an equation whose solution is 3

One of the equations could be:

$5x + 4 = 4x + 7$

Check

$5(3) + 4 = 4(3) + 7$

$15 + 4 = 12 + 7$

$19 = 19$

8) Isolate x

$7x - d = 6x + 2d$ We add "d" on both sides of the equation

$7x - d + d = 6x + 2d + d$

$7x = 6x + 3d$ We subtract $6x$ on both sides of the equation

$7x - 6x = 6x - 6x + 3d$

$x = 3d$

9) Isolate x

$2x - a = x - 3a + b$ We add "a" o both sides.

$2x - a + a = x - 3a + a + b$

$2x = x - 2a + b$ We subtract x on both sides

$2x - x = x - x - 2a + b$

$x = -2a + b$

10) Ten times a number minus 2 is 9 times the number plus "d" plus 5. Write the equation, solve it for x and check it.

$10x - 2 = 9x + d + 5$ We add 2 on both sides

$10x - 2 + 2 = 9x + 2 + d + 5$

$10x = 9x + d + 7$ We subtract $9x$ on both sides

$10x - 9x = 9x - 9x + d + 7$

$x = d + 7$

CHECK

$10(d + 7) - 2 = 9(d + 7) + d + 5$ We expand the brackets

$10d + 70 - 2 = 9d + 63 + d + 5$ We add or subtract the like terms.

$10d + 68 = 10d + 68$

$x = d + 7$ is the solution of the equation $10x - 2 = 9x + d + 5$

3.B. Solving Linear Equations

c. Solving two-step linear equations with multiplication and division

1) Solve: $5x - 2 = 4$

In this case, to isolate the variable x, we have to add 2 and then divide both sides of the equation with the same value 5.

$5x - 2 + 2 = 4 + 2$

$5x = 6$

$\frac{5}{5}x = \frac{6}{5}$

$x = \frac{6}{5}$

2) Solve and check $4x + 3 = 4 + x$

We will have to ISOLATE the unknown or the variable.

We will do the same operations in both terms so the equation is balanced all the time.

Step 1: minus 3 in both sides of the equation.

$4x + 3 - 3 = 4 - 3 + x$

$4x = 1 + x$

Step 2: subtract x in each side.

$4x - x = 1 + x - x$

$3x = 1$

Step 3: divide by 3 in both sides

$\frac{3x}{3} = \frac{1}{3}$

$x = \frac{1}{3}$

To be sure the result is correct, we check by substituting $x = \frac{3}{2}$ in the original equation.

CHECK

$4\left(\frac{1}{3}\right) + 3 = 4 + \frac{1}{3}$

$\frac{4}{3} + \frac{9}{3} = \frac{12}{3} + \frac{1}{3}$

$$\frac{13}{3} = \frac{13}{3}$$

Indeed, $x = \frac{1}{3}$ is the <u>solution</u> of the equation $4x + 3 = 4 + x$

3) Solve $\frac{5}{x} = 3 \ x \neq 0$

In this case, to isolate the variable x, we have to:

Step 1 multiply by x both sides of the equation

$$\frac{5}{x} * x = 3x$$

$$5 = 3x$$

Step 2 divide both sides of the equation with the same value 3.

Here we divide by 4

$$\frac{5}{3} = \frac{3x}{3}$$

$$x = \frac{5}{3}$$

4) Solve and check $3.2x + 5.2 = 9.4 - 2.1x$

We will have to ISOLATE the unknown or the variable.

We will do the same operations in both terms so the equation is balanced all the time.

Step 1: minus 5.2 in both sides of the equation.

$3.2x + 5.2 - 5.2 = 9.4 - 5.2 - 2.1x$

$3.2x = 4.2 - 2.1x$

Step 2: add $2.1x$ in both sides

$3.2x + 2.1x = 4.2 - 2.1x + 2.1x$

$5.3x = 4.2$

Step 3 divide by 5.3 in each side of the equation.

$$\frac{5.3x}{5.3} = \frac{4.2}{5.3}$$

$x = 0.79$

To be sure the result is correct, we check by substituting $x = 0.79$ in the original equation.

CHECK

$3.2(0.79) + 5.2 = 9.4 - 2.1(0.79)$

$2.54 + 5.2 = 7.74$

$7.74 = 7.74$

Indeed, $x = 0.79$ is the <u>solution</u> of the equation $3.2x + 5.2 = 9.4 - 2.1x$

5) Fifteen divided by a number is 5. Write then solve an equation to determine the number. Verify the solution.

$$\frac{15}{x} = 5$$

In this case, to isolate the variable x, we have to multiply by x and then divide both sides of the equation with the same value 5.

$$\frac{15x}{x} = 5x$$

$15 = 5x$

$\frac{15}{5} = \frac{5x}{5}$

$x = 3$

CHECK

$\frac{15}{3} = 5$

$5 = 5$

$x = 3$ is the solution of the equation $\frac{15}{x} = 5$

6) Solve $-6x = 8 - 22x$

We will have to ISOLATE the unknown or the variable.

We will do the same operations in both terms so the equation is balanced all the time.

Step 1: add $22x$ in both sides of the equation.

$-6x + 22x = 8 - 22x + 22x$

$16x = 8$

Step 2: divide by 16 in both sides

$\frac{16x}{16} = \frac{8}{16}$

$x = \frac{1}{2}$

7) Solve and check $13 - 5x = 4 - 4x$

We will have to ISOLATE the unknown or the variable.

We will do the same operations in both terms so the equation is balanced all the time.

Step 1: subtract 13 in both sides of the equation.

$13 - 13 - 5x = 4 - 13 - 4x$

$-5x = -9 - 4x$

Step 2: add $4x$ in both sides

$-5x + 4x = -9 - 4x + 4x$

$-x = -9$

Step 3: multiply with -1 in both sides of the equation

$-x * (-1) = -9 * (-1)$

$x = 9$

CHECK

$13 - 5(9) = 4 - 4(9)$

$13 - 45 = 4 - 36$

$-32 = -32$

So, $x = 9$ is the solution of the equation $13 - 5x = 4 - 4x$

8) Two rental halls are considered for a private concert.

Hall A costs $70 per person

Hall B costs $2500, plus $35 per person

Determine the number of people for which the halls will cost the same to rent.

Step 1 We model the problem with an equation.

Cost of Hall A is $70 times the number of people (x)

Cost of Hall B is $2485+$35 times the number of people (x)

The costs are equal, so the equation is:

$70x = 2485 + 35x$

Step 2 We find the number of people (x) by solving the equation.

$70x = 2485 + 35x$

We subtract $35x$ in both sides of the equation

$70x - 35x = 2485 + 35x - 35x$

$35x = 2485$

We divide by 35

$$\frac{35x}{35} = \frac{2485}{35}$$

$x = 71$ persons

Step 3 CHECK

$70(71) = 2485 + 35(71)$

$4970 = 2485 + 2485$

$4970 = 4970$

$x = 71$ persons, is the solution of the equation $70x = 2485 + 35x$

9) Seven subtract 4 times a number is equal to 5.3 times the same number, subtract 5. Write and solve the equation.

$7 - 4x = 5.3x - 5$

Step 1 Subtract 7 in both sides of the equation

$7 - 7 - 4x = 5.3x - 5 - 7$

$-4x = 5.3x - 12$

Step 2 Subtract $5.3x$ in both sides of the equation

$-4x - 5.3x = 5.3x - 5.3x - 12$

$-9.3x = -12$

Step 3 Divide by -9.3 each side

$$-\frac{9.3x}{-9.3} = -\frac{12}{-9.3}$$

$x = 1.29$

CHECK

$7 - 4(1.29) = 5.3(1.29) - 5$

$1.83=1.83$

$x = 1.29$ is the solution of the equation $7 - 4x = 5.3x - 5$

10) A car salesman is offered two methods of payment.

Plan A: $1400 per month with a commission of 5% on his sales.

Plan B: $1800 per month with a commission of 3% on his sales.

Sales are represented by the unknown x

a) Write an expression that will represent the total earnings using Plan A.

$E = 1400 + 0.05x$

b) Write an expression that will represent the total earnings using Plan B.

$E = 1800 + 0.03x$

c) Write an equation to determine the sales so the same total earnings are obtained from both plans.

$1400 + 0.05x = 1800 + 0.03x$

d) Solve the equation and explain what the answer represents.

$1400 + 0.05x = 1800 + 0.03x$

Step 1 subtract 1400 in each side of the equation

$1400 - 1400 + 0.05x = 1800 - 1400 + 0.03x$

$0.05x = 400 + 0.03x$

Step 2 Subtract $0.03x$ in each side of the equation

$0.05x - 0.03x = 400 + 0.03x - 0.03x$

$0.02x = 400$

Step 3 Divide by 0.02 in each side of the equation

$\frac{0.02x}{0.02} = \frac{400}{0.02}$

$x = \$20,000$

The value represents the total sales in dollars.

3.B. Solving Linear Equations

c. Solving two-step linear equations with distributivity property

1) Solve and check

$3(x + 2) = 7$

Step 1 we multiply 3 with x and 2 respectively

$3x + 6 = 7$

Step 2 we subtract 6 in both sides

$3x + 6 - 6 = 7 - 6$

$3x = 1$

Step 3 we divide by 3 in both sides

$\frac{3x}{3} = \frac{1}{3}$

$x = \frac{1}{3}$

CHECK

$3\left(\frac{1}{3} + 2\right) = 7$

$3 * \left(\frac{7}{3}\right) = 7$

$7 = 7$

$x = \frac{1}{3}$ is the solution of the equation $3(x + 2) = 7$

2) Solve and check

$5.2(x + 1.3) = 2.7$

Step 1 we multiply 5.2 with x and 1.3 respectively

$5.2x + 6.76 = 2.7$

Step 2 we subtract 6.76 in both sides

$5.2x + 6.76 - 6.76 = 2.7 - 6.76$

$5.2x = -4.06$

Step 3 we divide by 5.2 in both sides

$\frac{5.2x}{5.2} = \frac{-4.06}{5.2}$

$x = -0.78$

CHECK

$5.2(-0.78 + 1.3) = 2.7$

$5.2(0.51) = 2.7$

$2.7 = 2.7$

$x = -0.78$ is the solution of the equation

$5.2(x + 1.3) = 2.7$

3) Solve and check

$4(3x - 1) = 3(x + 5)$

Step 1 we multiply by 4 the first bracket and y 3 the second bracket

$12x - 4 = 3x + 15$

Step 2 we add 4 in both sides

$12x - 4 + 4 = 3x + 15 + 4$

$12x = 3x + 19$

Step 3 we subtract 3x in both sides

$12x - 3x = 3x - 3x + 19$

$9x = 19$

Step 4 we divide by 9 in both sides of the equation

$\frac{9x}{9} = \frac{19}{9}$

$x = 2.11$

CHECK

$4(3 * 2.11 - 1) = 3(2.11 + 5)$

$4(6.33 - 1) = 3(7.11)$

$4(5.33) = 21.33$

$21.33 = 21.33$

$x = 2.11$ is the solution of the equation $4(3x - 1) = 3(x + 5)$

4) Solve and check

$\frac{1}{2}(3x - 4) = \frac{3}{2}(2x + 5)$

Step 1 we multiply by 2 in both sides to get rid of the fractions

$\frac{2}{2}(3x - 4) = \frac{3*2}{2}(2x + 5)$

$3x - 4 = 3(2x + 5)$

Step 2 we multiply 3 with $2x$ and 5 respectively in the right side of the equation.

$3x - 4 = 6x + 15$

Step 3 we add 4 in both sides of the equation

$3x - 4 + 4 = 6x + 15 + 4$

$3x = 6x + 19$

Step 4 we subtract $6x$ in both sides

$3x - 6x = 6x - 6x + 19$

$-3x = 19$

Step 5 we divide by minus 3 each side

$-\frac{3x}{-3} = \frac{19}{-3}$

$x = -6.33$

CHECK

$\frac{1}{2}[3(-6.33) - 4] = \frac{3}{2}[2(-6.33) + 5]$

$\frac{1}{2}(-18.99 - 4) = \frac{3}{2}(-12.66 + 5)$

$\frac{1}{2}(-22.99) = 1.5(-7.66)$

$-11.49 = -11.49$

$x = -6.33$ is the solution for the equation $\frac{1}{2}(3x - 4) = \frac{3}{2}(2x + 5)$

5) Solve

$\frac{3}{2}(1 + 4x) = \frac{1}{3}(3 - 2x)$

Step 1 we multiply by the common denominator 6 each side

$\frac{3*6}{2}(1 + 4x) = \frac{1*6}{3}(3 - 2x)$

$9(1 + 4x) = 2(3 - 2x)$

Step 2 We apply the distributivity property in each side

$9 + 36x = 6 - 4x$

Step 3 we subtract 9 in each side of the equation

$9 - 9 + 36x = 6 - 9 - 4x$

$36x = -3 - 4x$

Step 4 we add $4x$ in each side

$36x + 4x = -3 - 4x + 4x$

$40x = -3$

Step 5 we divide by 40 each side

$$\frac{40x}{40} = -\frac{3}{40}$$

$$x = -\frac{3}{40}$$

6) Solve

$$\frac{x}{2} + \frac{x}{3} = x - \frac{1}{6}$$

Step 1 we multiply by 6 in both sides.

$$\frac{6x}{2} + \frac{6x}{3} = 6x - \frac{6}{6}$$

$$3x + 2x = 6x - 1$$

Step 2 we subtract $6x$ in both sides

$$3x + 2x - 6x = 6x - 6x - 1$$

$$-x = -1$$

Step 3 we multiply by -1 in each side.

$$x = 1$$

7) Solve

$$\frac{x}{3} + \frac{5}{3} = \frac{3}{4}$$

Step 1 we multiply by the common denominator 12 in each side

$$\frac{12x}{3} + \frac{12*5}{3} = \frac{12*3}{4}$$

$$4x + 20 = 9$$

Step 2 we subtract 20 in each side of the equation

$$4x + 20 - 20 = 9 - 20$$

$$4x = -11$$

Step 3 we divide by 4 in each side

$$\frac{4x}{4} = -\frac{11}{4}$$

$$x = -2\frac{3}{4}$$

8) Solve and check

$$3 - \frac{x}{12} = \frac{2x}{12} + 1$$

Step 1 we multiply by 12 (common denominator) all the terms in each side

$$3 * 12 - \frac{12x}{12} = \frac{2*12x}{12} + 1 * 12$$

$$36 - x = 2x + 12$$

Step 2 we subtract 36 in each side

$$36 - 36 - x = 2x + 12 - 36$$

$$-x = 2x - 24$$

Step 3 we subtract $2x$ in each side

$$-x - 2x = 2x - 2x - 24$$

$$-3x = -24$$

Step 4 we divide by -3 in each side

$$\frac{-3x}{-3} = \frac{-24}{-3}$$

$$x = 8$$

CHECK

$$3 - \frac{(8)}{12} = \frac{2*(8)}{12} + 1$$

$$3 - \frac{8}{12} = \frac{16}{12} + 1$$

$$3 - 0.66 = +1.34 + 1$$

$$2.34 = 2.34$$

$x = 8$ is the solution of the equation $3 - \frac{x}{12} = \frac{2x}{12} + 1$

9) Solve

$$\frac{5}{18} - \frac{x}{18} = \frac{3x}{6} + \frac{1}{2}$$

Step 1 we multiply by 18 all the terms in each side of the equation

$$\frac{5*18}{18} - \frac{x*18}{18} = \frac{3*18*x}{6} + \frac{18}{2}$$

$$5 - x = 9x + 9$$

Step 2 we subtract 5 in each side of the equation

$$5 - 5 - x = 9x + 9 - 5$$

$$-x = 9x + 4$$

Step 3 we subtract $9x$ in each side of the equation

$$-x - 9x = 4$$

$$-10x = 4$$

Step 4 we divide by -10 in each side of the equation

$$\frac{-10x}{-10} = \frac{4}{-10}$$

$$x = -0.4$$

10) Solve

$$\frac{1}{3}(2x - 3) + 4x - 3 = \frac{5}{6}(x + 1) + 2$$

Step 1 We multiply by 6 all terms in each side of the equation

$$\frac{1*6}{3}(2x - 3) + 4 * 6x - 3 * 6 = \frac{5*6}{6}(x + 1) + 2 * 6$$

$$2(2x - 3) + 24 - 18 = 5(x + 1) + 12$$

Step 2 we apply the distributive property for each bracket.

$$4x - 6 + 6 = 5x + 5 + 12$$

$$4x = 5x + 17$$

Step 3 we subtract $5x$ in each side of the equation

$$4x - 5x = 5x - 5x + 17$$

$$-x = 17$$

Step 4 we multiply by -1 in each side

$$-x * (-1) = 17 * (-1)$$

$x = -17$

3.C Equation of a straight line
a. Non-vertical and non-horizontal line

1) Correct

The equation of the line through M (-3,1) and slope -2 is $y = -2x - 5$

$m = \frac{y-y_1}{x-x_1}$

$-2 = \frac{y-1}{x-(-3)}$

$-2x - 6 = y - 1$

$y = -2x - 5$

2) Correct

The y intercept of the line $y = -2x - 5$ is y=-5

3) Incorrect

In the slope relation, $m = \frac{y-5}{x+4}$, the y intercept in terms of the slope m, is $b = 4m + 5$

$m = \frac{y-5}{x+4}$

$m(x + 4) = y - 5$

$mx + 4m = y - 5$

$mx + 5 + 4m = y$

$y = mx + 4m + 5$

$y = mx + b$

$b = 4m + 5$

4) Incorrect

The equation of the parallel line with $y = 3x + 1$ that passes through the point M (5,6) is $y = 3x - 9$

$m = \frac{y-6}{x-5} = 3$

$3(x - 5) = y - 6$

$3x - 15 = y - 6$

$y = 3x - 9$

5) Correct

Y intercept of the parallel line with $y = 3x + 1$ in problem 4 is b=-9

6) Incorrect

The intersection to x axis of $y = 4x - 8 \; is$ P (2,0)

$y = 0$

So,

$0 = 4x - 8$

$8 = 4x$

$x = \dfrac{8}{4} = 2$

7) Correct

The intersection to x axis of the line $y = -3x + 1$ is $P(\frac{1}{3}, 0)$

$0 = -3x + 1$

$-1 = -3x$

$x = \dfrac{1}{3}$

8) Incorrect

The slope of the line with x intercept = 4 and passing through M(2,3) is $m = \dfrac{-3}{2}$

Slope is $m = \dfrac{y - y_1}{x - x_1}$

$m = \dfrac{y - 3}{x - 2}$

We know that, the x intercept has y=0. In our case, x intercept has the coordinates (4,0)

$m = \dfrac{0 - 3}{4 - 2} = \dfrac{-3}{2}$

9) Correct

The slope of the line that has x intercept the point P(2,0) and passes through the point M(6,1. is m = - 1/4

Slope is $m = \dfrac{y - y_1}{x - x_1}$

$m = \dfrac{y - 1}{x - 6}$

We know that, the x intercept has y=0. In our case, x intercept has the coordinates (2,0)

$m = \dfrac{0 - 1}{2 - 6} = \dfrac{-1}{4}$

10) Incorrect

The slope of line BC is

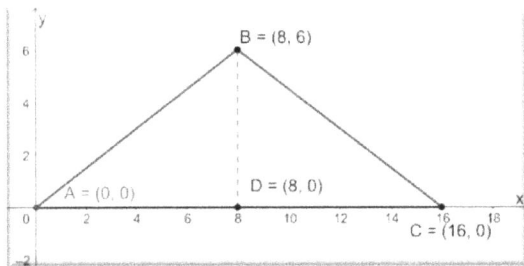

$m = \dfrac{y_2 - y_1}{x_2 - x_1} = \dfrac{6 - 0}{8 - 16} = \dfrac{6}{-8} = \dfrac{-3}{4}$

11) Determine the slope of each line

a) $y = 3x - 4$ b) $y = -2x + 3$ c) $y = 4x - 7$

a)

The slope is the number that is multiplied with the variable x. In this case $slope = 3$

b)

The slope is the number that is multiplied with the variable x. In this case $slope = -2$

c)

The slope is the number that is multiplied with the variable x. In this case $slope = 4$

12) Graph the line $2(x - 3) = (y + 1)$

We need to transform this equation in the slope and y-intercept form $y = mx + b$. This is because the easiest form of a line that we can graph is the slope and y-intercept form

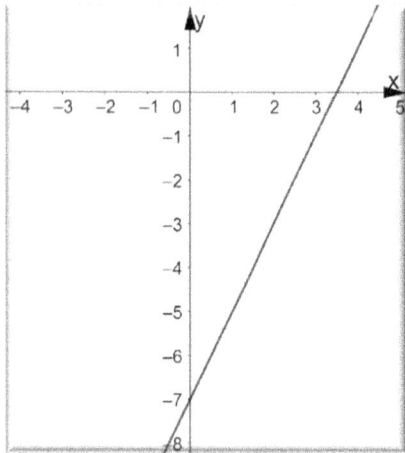

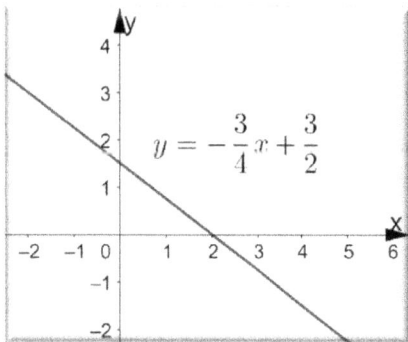

Step 1 We apply the distributivity property to the first bracket.

$2x - 6 = y + 1$

Step 2 we subtract 1 in each side of the equation

$2x - 6 - 1 = y + 1 - 1$

$2x - 7 = y$

Or

$y = 2x - 7$

13) Determine the slope and graph the line: $3x + 4y - 6 = 0$

We need to transform this equation in the slope and y-intercept form $y = mx + b$. This is because the easiest form of a line that we can graph is the slope and y-intercept form.

Step 1 we subtract $3x$ in both sides of the equation.

$3x - 3x + 4y - 6 = -3x$

$4y - 6 = -3x$

Step 2 we add 6 in both sides of the equation

$4y - 6 + 6 = -3x + 6$

$4y = -3x + 6$

Step 3 we divide by 4 in each side of the equation

$\frac{4}{4}y = -\frac{3}{4}x + \frac{6}{4}$

$y = -\frac{3}{4}x + \frac{3}{2}$

14) Plot the following equations on the same graph. What do you notice?

a) $5x + 3y - 2 = 0$ b) $3x + 3y - 2 = 0$ c) $x + 3y - 2 = 0$

a) Step 1 we add 2 in each side of the equation

$5x + 3y - 2 + 2 = 2$

$5x + 3y = 2$

Step 2 we subtract $5x$ in each side of the equation

$5x - 5x + 3y = 2 - 5x$

$3y = -5x + 2$

Step 3 we divide by 3 in each side

$$\frac{3y}{3} = -\frac{5}{3}x + \frac{2}{3}$$

$$y = -\frac{5}{3}x + \frac{2}{3}$$

b)

Step 1 a) Step 1 we add 2 in each side of the equation

$3x + 3y - 2 + 2 = 2$

$3x + 3y = 2$

Step 2 we subtract $3x$ in each side of the equation

$3x - 3x + 3y = 2 - 3x$

$3y = -3x + 2$

Step 3 we divide by 3 in each side

$$\frac{3y}{3} = -\frac{3}{3}x + \frac{2}{3}$$

$$y = -x + \frac{2}{3}$$

c) Step 1 we add 2 in each side of the equation

$x + 3y - 2 + 2 = 2$

$x + 3y = 2$

Step 2 we subtract x in each side of the equation

$x - x + 3y = 2 - x$

$3y = -x + 2$

Step 3 we divide by 3 in each side

$$\frac{3y}{3} = -\frac{1}{3}x + \frac{2}{3}$$

$$y = -\frac{1}{3}x + \frac{2}{3}$$

We notice that the graphs intersect the y axis in the same point $\frac{2}{3}$ but they have different slopes.

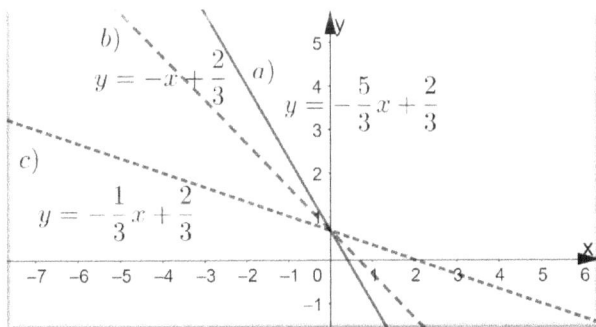

3.D. Straight-line graph

1) Correct

The slope of the line $y = 3x - 1$ is

2) Incorrect

The y intercept of the line $2x + 3y = 5$ is $b = \frac{5}{3}$

$2x + 3y = 5$

$3y = -2x + 5$

$y = -\frac{2}{3}x + \frac{5}{3}$

3) Correct

The slope of the line $-3x + 5y = -1$ is $m = \frac{3}{5}$

$5y = 3x - 1$

$y = \frac{3}{5}x - \frac{1}{5}$

4) Incorrect

The y intercept of the line $2(x - 5) + 3(y + 2) = 2$ is $b = 2$

$2(x - 5) + 3(y + 2) = 2$

$2x - 10 + 3y + 6 = 2$

$2x + 3y - 4 = 2$

$3y = -2x + 6$

$y = -\frac{2}{3}x + 2$

5) Correct

The slope of the line $-3(x + 2) - 4(y - 7) = 6$ is $m = -\frac{3}{4}$

$-3(x + 2) - 4(y - 7) = 6$

$-3x - 6 - 4y + 28 = 6$

$-3x - 4y + 22 = 6$

$-4y = 3x - 16$

$4y = -3x + 16$

$y = -\frac{3}{4}x + 4$

The next problems use the figure shown below

6) Incorrect

The line a) has the equation
Y=2x+3

7) Incorrect

The line b) has the equation
Y= 8x-4

8) Correct

The line c) has the equation

$Y = -\frac{1}{8}x + 1$

9) Incorrect

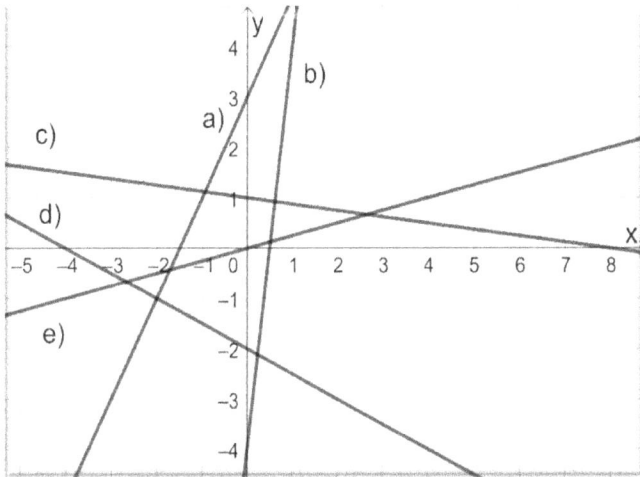

The line d) has the equation

$y = -\frac{1}{2}x - 2$

10) Correct

The line e) has the equation

$y = \frac{1}{4}x$

3.E. Special cases of linear equations:
Vertical and horizontal lines

1) The equation of horizontal line through M (3,4) is $y = 4$

Problems 2,3,4,5 will be based on the figure shown below.

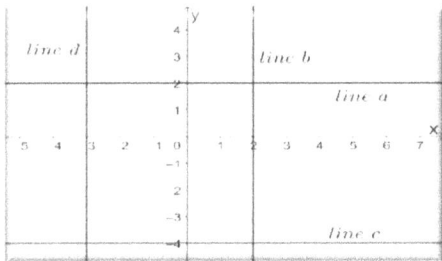

2) The equation of line a is $y = 2$

3) The equation of line b is $x = 3$

4) The equation of line c is $y = -4$

5) The equation of line d is $x = -3$

Problems 6 and 7 will be based on the figure shown below.

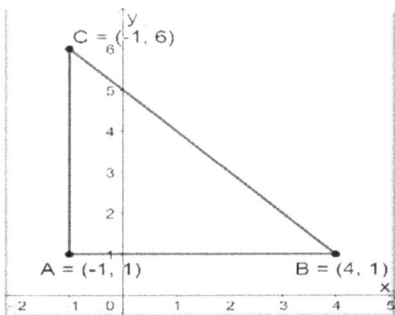

6) The equation of the line that passes through the points C and A in the figure below is $x = -1$

7) The equation of the line that passes through the points A and B in the figure below is $x = 1$

Problems 8 and 9 will be based on the figure shown below.

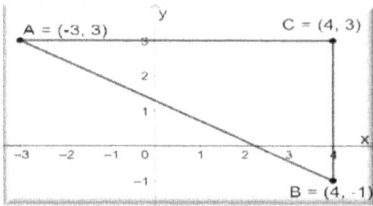

8) The equation of the segment AC is $y = 3$

9) The equation of the segment CB is $x = 4$

Problem 10 will be based on the figure shown below.

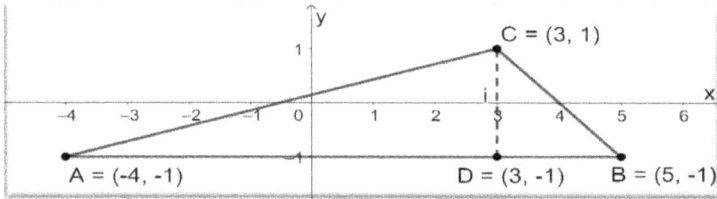

10) The equation of the segment CD is $x = 3$

3.F Parallel and perpendicular lines
a. Parallel lines

1) Correct

The equation of the parallel line with $y = x - 1$ that intersects y axis at point M $(0,5)$ is $y = x + 5$

2) Incorrect

The equation of the parallel line with $y = -3x + 2$ that intersects y axis at point M $(0,-3)$ is $y = -3x - 3$

3) Correct

The line $y = -5x + 3$ is not parallel with $y = -4x + 3$

4) Incorrect

The $y = 3x - 1$ is the same with $y = 3x - 1$

5) Incorrect

The line $y = 4x + 3$ is parallel with $y = 4x - 25$

Problems 6 and 7 will be based on the figure shown below.

6) Correct

The line a is parallel to line b. $m_a = m_b = \frac{3}{2}$

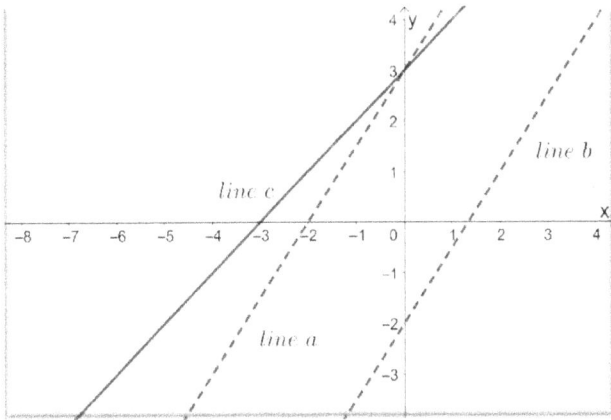

7) Incorrect

The equation of the line a is $y = \frac{3}{2}x + 3$

8) Correct

The equation of the line b is $y = \frac{3}{2}x - 2$

9) Incorrect

Line c is not parallel with line a and line b

10) Correct

The equation of the line c is $= x + 3$

3.F Parallel and perpendicular lines
b. Perpendicular lines

1) Correct

The lines that have the equations $y = 3x + 4$ and $y = -\frac{1}{3}x - 5$ are perpendicular.

2) Correct

The lines that have the equations $y = 3x - 3$ and $y = 3x + 3$ are not perpendicular.

3) Incorrect

The line perpendicular to the line $y = -2x + 7$ has the slope $m = \frac{1}{2}$.

4) Correct

The equation of the line perpendicular to $y = 5x - 1$ that passes through M (3,4) is

$y = -\frac{1}{5}x + 4.6$

$m = -\frac{1}{5}$

The perpendicular line goes through M (3,4)

$4 = -\frac{1}{5}(3) + b$

$4 + \frac{1}{5}(3) = b$

$b = \frac{20+3}{5} = \frac{23}{5} = 4.6$

$y = -\frac{1}{5}x + 4.6$

5) Incorrect

The equation of the line perpendicular to $y = -2x + 3$ through M (-2,-3) is

$y = \frac{1}{2}x - 2$

The slope of the perpendicular line is $m = \frac{1}{2}$

We substitute the coordinates of M into the formula $y = mx + b$

$-3 = \frac{1}{2}(-2) + b$

$-3 = -1 + b$

$-3 + 1 = +b$

$b = -2$

So, the equation of the line perpendicular to $y = -2x + 3$ through M (-2,-3) is

$y = \frac{1}{2}x - 2$

6) Correct

The equation of the line perpendicular to $y = x + 2$ through K (1,2) intersects the y axis in M (0,3)

The slope of the perpendicular to $y = x + 2$ will be $m = -1$

We substitute the coordinates of K into the formula $y = mx + b$

$2 = (-1)(1) + b$

$2 + 1 = b$

$b = 3$

So, the equation of the line perpendicular to $y = x + 2$ through K (1,2) is $y = -x + 3$

Y intercept will be 3

7) Incorrect

The equation of the line perpendicular to AB in point B, is $y = \frac{7}{3}x - 11.33$

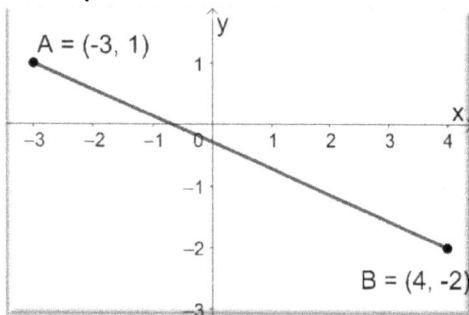
A = (-3, 1)
B = (4, -2)

Slope of AB is $m = \frac{y_2 - y_1}{x_2 - x_1} = \frac{-2-1}{4-(-3)} = \frac{-3}{7}$

So,

The slope of the perpendicular to AB is $m = \frac{7}{3}$

The perpendicular goes through B so, we substitute point B coordinates into $y = mx + b$

$-2 = \frac{7}{3}(4) + b$

$-2 - \frac{28}{3} = b$

$b = -11.33$

So, the equation of the line perpendicular to AB through B (4,-2) is $y = \frac{7}{3}x - 11.33$

8) Incorrect

The equation of the line perpendicular to AB in problem 7 will intersect x axis in $x =$

The equation of the line perpendicular to AB through B (4,-2) is $y = \frac{7}{3}x - 11.33$.

The x intercept will have the y coordinate equal zero.

$0 = \frac{7}{3}x - 11.33$

$\frac{7}{3}x = 11.33$

$x = \frac{11.33*3}{7} = \frac{33.99}{7} = 4.84$

9) Correct

The perpendicular to AB through the point B (6,1) will intersect y axis in $b = 13$

The slope of AB is $m = \frac{y_2-y_1}{x_2-x_1} = \frac{1-(-3)}{6-(-2)} = \frac{4}{8} = \frac{1}{2}$

So, the slope of the perpendicular to AB will be $m = -2$

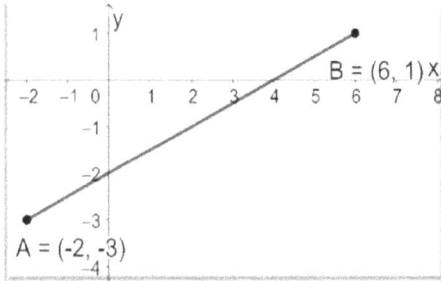

The perpendicular goes through B(6,1) so, we substitute point B coordinates into $y = mx + b$

$1 = -2(6) + b$

$1 + 12 = b$

$b = 13$

10) Incorrect

The perpendicular to AB through the point B (6,1) will intersect x axis at $x = 6.5$

We know that for x intercept y coordinate will be zero.

$0 = -2x + 13$

$2x = 13$

$x = \frac{13}{2} = 6.5$

4.A Express linear inequalities graphically and algebraically

1) Represent on the number line and algebraically:

A number bigger than and equal to -5

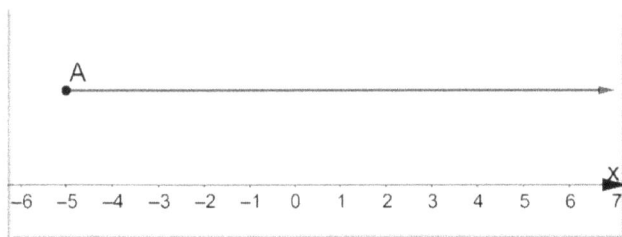

$x \geq -5$

2) Represent on the number line and algebraically:

A number bigger than and equal to 1

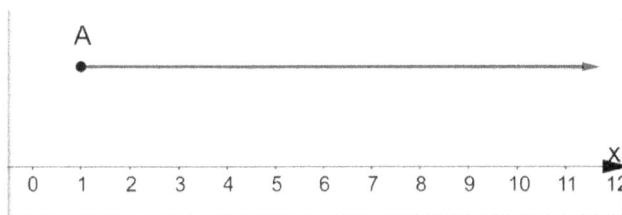

$x \geq 1$

3) Represent on the number line and algebraically:

A number less than and equal to 2

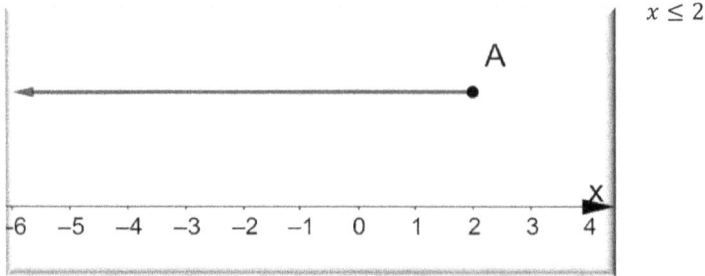

$x \leq 2$

A

-6 -5 -4 -3 -2 -1 0 1 2 3 4

4) Is each inequality true or false?

a) $3 < 7$ b) $5 \leq -8$ c) $\frac{1}{5} < \frac{1}{7}$

a) $3 < 7$ true   b) $5 \leq -8$ 5 is bigger than -8

c) $\frac{1}{5} < \frac{1}{7}$

To check which fraction is grater we transform each one into equivalent fractions with the same denominator.

$\frac{1*7}{5*7} < \frac{1*5}{7*5}$

$\frac{7}{35} > \frac{5}{35}$

So,

$\frac{1}{5} < \frac{1}{7}$ is false

5) Use a symbol $>, <, \leq, or \geq$ to write an inequality that corresponds to each statement.

a) x is less than -7

$x < -7$

b) a number is greater and equal than 3

$x \geq 3$

c) x is negative

$x < 0$

6) Is each number a solution of $x \leq -3$?

a) 0   b) 5   c) -7

a) 0   b) 5   c) -7

We represent the solutions on the number line.

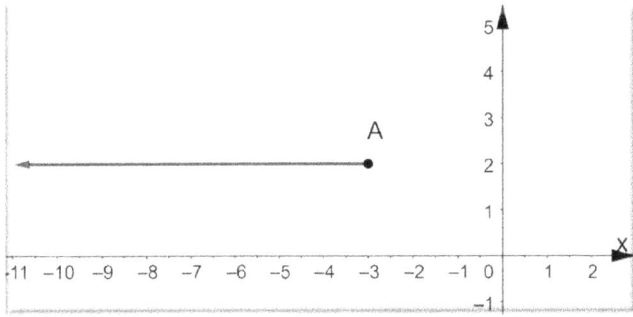

We can see that value 0 and 5 are outside of the solution. Only the value of -7 is part of all the solutions.

7) Write 3 numbers that are solutions of each inequality.

a) $a > 3$ b) $b \leq 3$ c) $w < -5$

a) $c > 3$ b) $d \leq 3$ c) $w < -5$
4,5,6   -3, 0, 3   -10, -7.5, -4.99

8) Determine whether the given number is a solution.

a) $y < 2, 2$ b) $x > 4, 0$ c) $z \leq 5, 0$

a) $y < 2,$ b) $x > 4, 0$
two is not part of the solution   zero is less than four
                                    not a solution

c) $z \leq 5,$ all the numbers have to be smaller than 5, including 5. Zero is smaller than 5, so it is part of the solutions.

9) Iveta and Emma write the inequality whose solution is shown below.

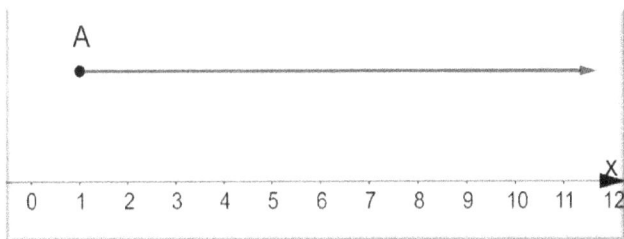

Iveta writes $x \geq 1$
Emma writes $1 > x$
Who is correct?

The solution are all the points that are bigger and equal with 1.
Iveta is correct, Emma is wrong (she wrote $x < 1$

10) Graph the solution for $x \leq -2$

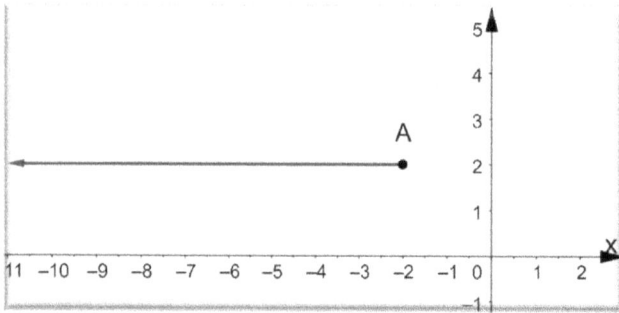

4.B Solving one-step linear inequalities

1) Solve and graph the solution on the number line.

$x - 7 \geq -4$

We add 7 in each side.

$x - 7 + 7 \geq -4 + 7$

$x \geq 3$

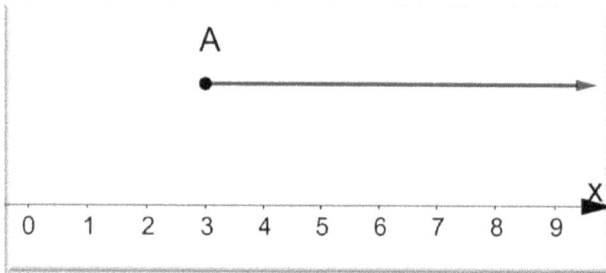

2) Which operation will you perform on each side of the inequality to isolate the variable?

a) $y + 3 > 5$ b) $k - 3.4 < 7.8$ c) $5.6 < 2.1 + n$

a) subtract 3   b) add 3.4   c) subtract 2.1

3) What must be done to the first inequality to get to the second inequality?

a) $x - 5 \leq 4$ b) $2x \leq 4$ c) $x - \frac{1}{3} > \frac{2}{3}$

$x \leq 9$ $x \leq 2$ $x > 1$

a) add 5 on each side   divide by 2   add $\frac{1}{3}$

4) State three values that satisfy each inequality; one integer, one fraction, and one decimal

a) $x + 3 < 5$

b) $x - 4 > 1$

c) $5x \leq 20$

a) subtract 3 on each side

$x + 3 - 3 < 5 - 3$

$x < 2$

Three of the solutions are: $-1, \frac{1}{2}, 1.89$

b) add 4 on each side

$x - 4 + 4 > 1 + 4$

$x > 5$

Three of the solutions are: $10, 7\frac{1}{2}, 5.89$

c) divide by 5 on both sides

$\frac{5}{5}x \leq \frac{20}{5}$

$x \leq 4$

Three of the solutions are: $-15.34, \frac{1}{2}, 4$

5) Solve, graph and check the inequality:

$4x \leq 12$

Divide by 4 on each side

$\frac{4}{4}x \leq \frac{12}{4}$

$x \leq 3$

Graph

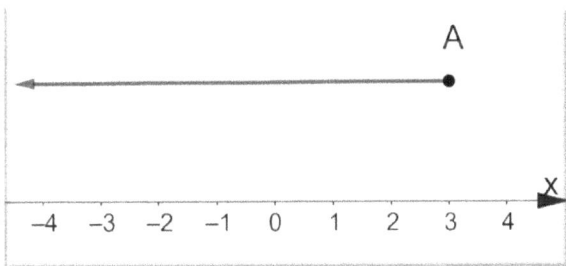

CHECK

We choose one value that is smaller that 3 and substitute it into the original inequality.

Let us choose one.

So,

$4(1) \leq 12$

$4 \leq 12$ is true.

The solution are all the real numbers less or equal with 3.

6) Is 2 a solution of the inequalities below?

a) $x - 3 > -4$ 　　　　　　　b) $y + 4 \leq 5$ 　　　　　　　c) $m + 2 < 6$

a) $x - 3 + 3 > -4 + 3$
$x > -1$
2 is part of the solution of the inequality $x - 3 > -4$

b) $y + 4 \leq 5$
$y + 4 - 4 \leq 5 - 4$
$y \leq 1$
2 is not part of the solution of the inequality $y + 4 \leq 5$

c) $m + 2 < 6$
$m + 2 - 2 < 6 - 2$
$m < 4$
2 is part of the solution of the inequality $m + 2 < 6$

7) Melissa has $310 in her bank account. She must maintain a minimum balance of $600 in her bank account to avoid paying a monthly fee. How much money can Melissa deposit into her account to avoid paying bank fees?
a) Choose a variable and write an inequality to solve the problem.
b) Solve the problem

a) The variable will be x – the money to be deposited, and the inequality would be:
$310 + x \geq 600$

b) $310 + x \geq 600$
$310 - 310 + x \geq 600 - 310$
$x \geq 290$
So, Melissa has to deposit at least $290 in order to avoid paying bank fees.

8) Mark is saving money to buy a camera for the next camping trip. He earned $200 during the weekends, but he still did not have the $750 he needed for the camera.
a) Choose a variable for the money needed, then write an inequality to represent this situation.
b) Solve the inequality.
c) Verify the solution.

a) The variable could be x. The inequality is:
$200 + x \geq 750$

b) Solve
$200 - 200 + x \geq 750 - 200$

$x \geq 550$

c) Verify

We choose any value that is bigger than 550, say 600

So,

$200 + 600 \geq 750$

$800 \geq 750$

So, any value that is bigger or equal with \$550 is a solution.

9) Write and solve an inequality to show how many cars Samuel has to wash at \$7 a car to earn at least \$350?

$7x \geq 350$

$\frac{7}{7}x \geq \frac{350}{7}$

$x \geq 50$ cars

10) A water slide charges \$2 to rent an inflatable ring, and \$0.5 per ride. Iveta has \$12. How many rides can Iveta go on?

$2 + 0.5x \geq 12$

$2 - 2 + 0.5x \geq 12 - 2$

$0.5x \geq 10$

$\frac{0.5}{0.5}x \geq \frac{10}{0.5}$

$x \geq 20$ rides

4.C Solving multi-step linear inequalities

1) Solve and check

$5x + 7 \geq 2$

Step 1 we subtract 7 on each side of the inequality.

$5x + 7 - 7 \geq 2 - 7$

$5x \geq -5$

Step 2 we divide by 5 on each side of the inequality

$\frac{5}{5}x \geq -\frac{5}{5}$

$x \geq -1$

CHECK

We choose any number that is greater or equal with minus 1

Let us choose 0 and substitute it instead of x in the original inequality

$5(0) + 7 \geq 2$

$0 + 7 \geq 2$

$7 \geq 2$ which is true

So, $x \geq -1$ is the solution for the inequality $5x + 7 \geq 2$

2) Solve and graph the solution

$3x + 4 \geq 6 + 2x$

Step 1 we subtract 4 on each side of the inequality

$3x + 4 - 4 \geq 6 - 4 + 2x$

$3x \geq 2 + 2x$

Step 2 we subtract $2x$ on both sides.

$3x - 2x \geq 2 + 2x - 2x$

$x \geq 2$

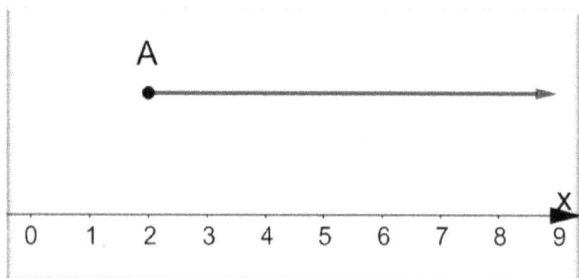

3) Solve

$6.2x + 3.1 < 2.3x + 1.3$

Step 1 we subtract 3.1 on each side

$6.2x + 3.1 - 3.1 < 2.3x + 1.3 - 3.1$

$6.2x < 2.3x - 1.8$

Step 2 we subtract $2.3x$ on both sides

$6.2x - 2.3x < 2.3x - 2.3x - 1.8$

$3.9x < 1.0$

Step 3 we divide by 3.9 on each side

$\frac{3.9}{3.9}x < -\frac{1.8}{3.9}$

$x < -0.46$

4) Solve and check

$1 + x > 3 + \frac{1}{3}x$

Step 1 we subtract 1 on each side of the inequality

$1 - 1 + x > 3 - 1 + \frac{1}{3}x$

$x > 2 + \frac{1}{3}x$

Step 2 we subtract $\frac{1}{3}x$ on each side

$x - \frac{1}{3}x > 2 + \frac{1}{3}x - \frac{1}{3}x$

$x - \frac{1}{3}x > 2$

Step 4 we take common factor x on the right side of the inequality

$x\left(1 - \frac{1}{3}\right) > 2$

Step 5 we subtract 1/3 from 1 in the bracket

$x\left(\frac{3}{3} - \frac{1}{3}\right) > 2$

$\frac{2}{3}x > 2$

Step 6 we multiply by 3/2 on both sides

$\frac{2}{3} * \frac{3}{2}x > 2 * \frac{3}{2}$

$x > 3$

CHECK

We choose the value 6 to be substituted instead of x in the original inequality

$1 + 6 > 3 + \frac{1}{3}(6)$

$7 > 3 + 2$

7>5 is true

So,

$x > 3$ is the solution of the inequality $1 + x > 3 + \frac{1}{3}x$

5) Solve

$\frac{2}{5}x - \frac{1}{2} > 3 + x$

Step 1 add ½ on each side of the inequality

$\frac{2}{5}x - \frac{1}{2} + \frac{1}{2} > 3 + \frac{1}{2} + x$

$\frac{2}{5}x > \frac{7}{2} + x$

Step 2 subtract x on each side of the inequality

$\frac{2}{5}x - x > \frac{7}{2} + x - x$

$x\left(\frac{2}{5} - 1\right) > \frac{7}{2}$

Step 3 we follow the order of operations rule and do the bracket $\left(\frac{2}{5} - 1\right)$

$x\left(\frac{2}{5} - \frac{5}{5}\right) > \frac{7}{2}$

$-\frac{3}{5}x > \frac{7}{2}$

Step 4 we multiply by 10 (common denominator) on each side of the inequality

$-\frac{3*10}{5}x > \frac{7*10}{2}$

$-3 * 2x > 7 * 5$

$-6x > 35$

Step 5 we multiply by -1 on each side of the inequality

The sign of the inequality changes in <

We have:

$6x < -35$

Step 6 we divide by 6 on each side of the inequality

$\frac{6}{6}x < -\frac{35}{6}$

$x < -5\frac{5}{6}$

6) Your school wants to raise money for charity. The school organizes a dance where the DJ costs $1200 and the ticket costs $8. How many tickets have to be sold to make a profit more than $1700?
a) Write an inequality to solve the problem
b) Solve and verify the solution

a) The variable x represents the number of students that buy tickets.
The inequality is:

$8x - 1200 \geq 1700$

b) Solve

Step 1 add 1200 on both sides of the inequality

$8x - 1200 + 1200 \geq 1700 + 1200$

$8x \geq 2900$

Step 2 we divide by 8 on both sides of the inequality

$\frac{8}{8}x \geq \frac{2900}{8}$

$x \geq 362$ students

CHECK

We choose a value of 400 for x

$8(400) - 1200 \geq 1700$

$3200 - 1200 \geq 1700$

$2000 \geq 1700$ it is true

So, $x \geq 362$ students is the solution for the inequality $8x - 1200 \geq 1700$

7) Solve

$7 + \frac{1}{3}x > 2(x + 12)$

Step 1 we apply the distributivity property and expand the bracket

$7 + \frac{1}{3}x > 2x + 24$

Step 2 we subtract 7 on both sides of the inequality

$7 - 7 + \frac{1}{3}x > 2x + 24 - 7$

$\frac{1}{3}x > 2x + 17$

Step 3 we subtract $2x$ on both sides

$\frac{1}{3}x - 2x > 2x - 2x + 17$

$\frac{1}{3}x - 2x > 17$

Step 4 we subtract the like terms on the left side of the inequality

$-\frac{5}{3}x > 17$

Step 5 we multiply by -1 on both sides. The sign of the inequality will flip to <

$-\frac{5}{3}(-1)x > 17(-1)$

$\frac{5}{3}x < -17$

Step 6 we multiply with 3/5 on both sides of the inequality

$\frac{5}{3} * \frac{3}{5}x < -17 * \frac{3}{5}$

$x < -\frac{51}{5}$

$x < -10\frac{1}{5}$

8) Solve

$3(x - 3) > \frac{2}{3}(3x + 6)$

Step 1 we expand the rackets by using the **distributivity property**

$3x - 9 > 2x + 4$

Step 2 we add 9 on both sides of the inequality

$3x - 9 + 9 > 2x + 4 + 9$

$3x > 2x + 13$

Step 3 we subtract $2x$ on each side of the inequality

$3x - 2x > 2x - 2x + 13$

$x > 13$

9) Solve

$\frac{1}{2}x + \frac{5}{3} \leq \frac{3}{2}x - \frac{1}{4}$

Step 1 we multiply y 12 (common denominator) on each side

$\frac{1*12}{2}x + \frac{5*12}{3} \leq \frac{3*12}{2}x - \frac{1*12}{4}$

$6x + 20 \leq 18x - 3$

Step 2 we subtract 20 on each side of the inequality

$6x + 20 - 20 \leq 18x - 3 - 20$

$6x \leq 18x - 23$

Step 3 we subtract $18x$ on each side of the inequality

$6x - 18x \leq 18x - 18x - 23$

$-12x \leq -23$

Step 4 we multiply by -1 on each side of the inequality

$-12(-1)x \leq -23(-1)$

The sign of the inequality will flip to $\geq$

$12x \geq 23$

Step 5 we divide y 12 on each side of the inequality

$\frac{12}{12}x \geq \frac{23}{12}$

$x \geq 1\frac{11}{12}$

10) John is replacing the light bulbs in his house from regular to energy saver light bulbs.
A regular light bulb costs $0.6 and has an electricity cost of $0.005 per hour.
An energy saver light bulb costs $5.5 and has an electricity cost of $0.001 per hour.
For how many hours of use it is cheaper to use an energy saver light bulb than a regular light bulb?
a) Write an inequality for this problem.
b) Solve the inequality. Explain the solution in words.

a) $5.5 + 0.001x < 0.6 + 0.004x$

b) Solve it

Step 1 we subtract 5.5 on each side of the inequality

$5.5 - 5.5 + 0.001x < 0.6 - 5.5 + 0.004x$

$0.001x < -4.9 + 0.004x$

Step 2 we subtract $0.004x$ on each side of the inequality

$0.001x - 0.004x < -4.9 + 0.004x - 0.004x$

$-0.003x < -4.9$

Step 3 we multiply by -1 on each side of the inequality

$-0.003(-1)x < -4.9(-1)$

The sign of the inequality changes to $>$

$0.003x > 4.9$

Step 4 we divide by 0.003 on each side of the inequality

$\frac{0.003}{0.003}x > \frac{4.9}{0.003}$

$x > 1,633.3 \ hours$

The energy saver light bulb has to be used more than 1,633.3 hours to be cost effective

4.D. Linear inequalities with two variables

1) Graph the solution

$y > 3x - 1$

Step 1 we graph the line $y = 3x - 1$

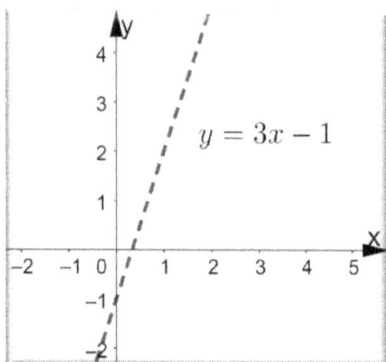

Step 2 we choose a test point, say (0,0) and see if this point satisfies the inequality.

$y > 3x - 1$

So,

$0 > 3(0) - 1$

$0 > 0 - 1$

$0 > -1$ it is true

It means that point (0,0) is part of the solution. The solution is all the points situated at the left of the line $y = 3x - 1$

We will graph the solution.

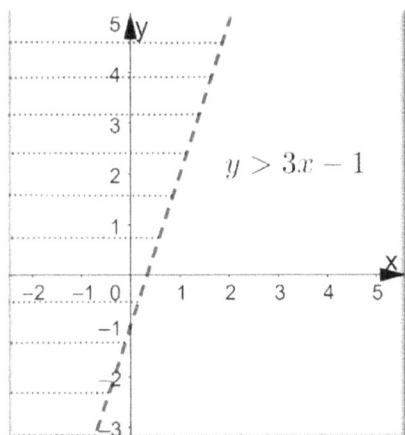

$y > 3x - 1$

2) Which point(s) is/are in the solution region of the inequality $3x - 6y \leq 4$

a) (0,0)   b) (4,2)   c) (-2,5)   d) (3,-6)

a) we substitute the coordinates (0,0) in the inequality
$3(0) - 6(0) \leq 4$
$0 \leq 4$ is true, so, point (0,0) is in the solution region of the inequality $3x - 6y \leq 4$
b) we substitute the coordinates (4,2) in the inequality
$3(4) - 6(2) \leq 4$
$0 \leq 4$ is true, so, point (4,2) is in the solution region of the

inequality $3x - 6y \leq 4$
c) we substitute the coordinates (-2,5) in the inequality
$3(-2) - 6(5) \leq 4$
$-6 - 30 \leq 4$
$-36 \leq 4$
is true, so, point (-2,5) is in the solution region of the inequality $3x - 6y \leq 4$
d) we substitute the coordinates (3,-6) in the inequality
$3(3) - 6(-6) \leq 4$
$9 + 36 \leq 4$
$45 \leq 4$
is not true, so, point (3,-6) is not in the solution region of the inequality $3x - 6y \leq 4$

3) Sketch the graphs of the following inequalities.
a) $y \geq 0$ b) $x > 2 \ and \ y > 1$

a)

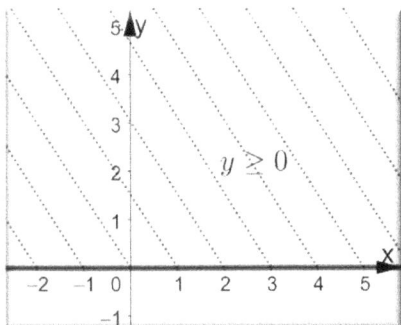

$y \geq 0$

b)

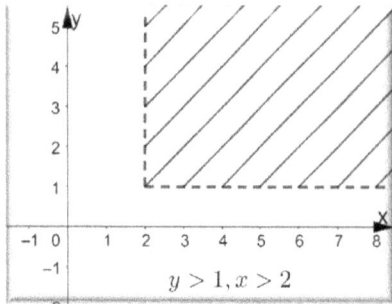

$y > 1, x > 2$

4) Graph the solution of two variable inequality

$4x + 3y \geq -9$

The idea here is to transform this inequality in something like

$y \geq mx \pm b$

Step 1 we subtract $4x$ on both sides of the inequality

$4x - 4x + 3y \geq -9 - 4x$

$3y \geq -4x - 9$

Step 2 we divide by 3 on each side of the inequality

$\frac{3}{3}y \geq -\frac{4}{3}x - \frac{9}{3}$

$y \geq -\frac{4}{3}x - 3$

Step 3 we graph the line $y = -\frac{4}{3}x - 3$

Because the inequality has the sign $\geq$ it means that the values on the line are part of the solution.

We represent the line as a full line not dotted line.

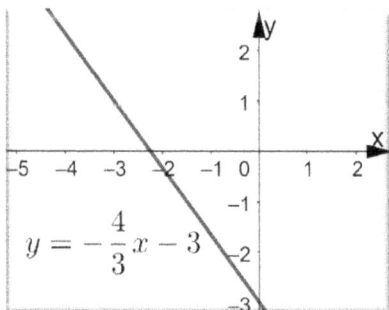

$y = -\frac{4}{3}x - 3$

Step 2 we choose the test point (0,0) to check if this test point is part of the solution

We substitute the coordinates 0,0 instead of x and y in the inequality

$0 \geq -\frac{4}{3}(0) - 3$

$0 \geq -3$ which is true,

So, the solution is the region above the line, including the points on the line

$y = -\frac{4}{3}x - 3$

Step 3 we graph the solution.

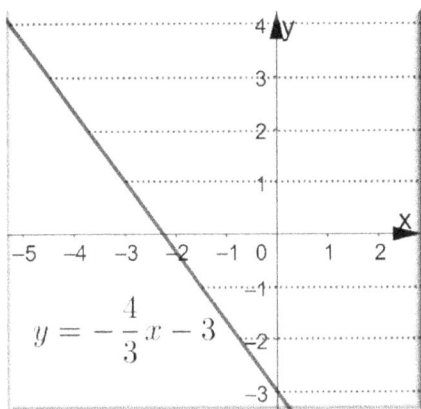

$$y = -\frac{4}{3}x - 3$$

5) The graph below, the equation of the boundary line is: $x - 3y = 6$

Determine the inequality represented by the graph.

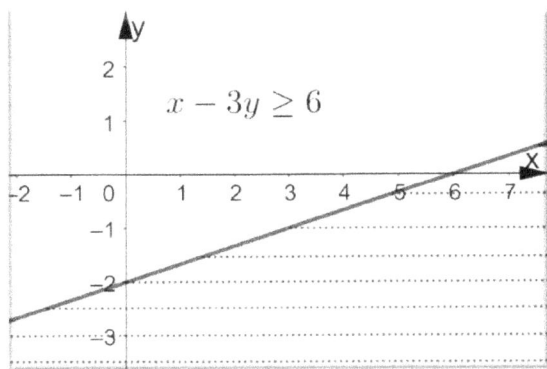

$x - 3y \geq 6$

Step 1 we add $3y$ on both sides of the equation

$x - 3y = 6$

$x - 3y + 3y = 6 + 3y$

$x = 6 + 3y$

Step 2 we subtract 6 on both sides of the equation

$x - 6 = 6 - 6 + 3y$

$x - 6 = 3y$

Step 3 we divide by 3 on both sides of the equation

$\frac{1}{3}x - \frac{6}{3} = \frac{3}{3}y$

$\frac{1}{3}x - 2 = y$

$y = \frac{1}{3}x - 2$

Step 4 we substitute point (0,0) into the equation and use the sign $\leq$ *or* $\geq$ in such a way that the point is not part of the solution.

$0 = \frac{1}{3}(0) - 2$

$0 = -2$

So, we will use the sign that will not satisfy the inequality for (0,0)

$x - 3y \geq 6$

6) The graph below shows the solution to the inequality

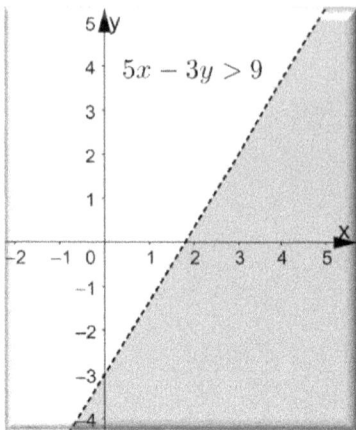

a) Explain why the boundary line is a broken line?

The left side is only bigger than the right side, not equal.

The points situated on the line are not part of the solution. To show that, the line is represented boken and not continue.

b) Why the solution is beneath the line and not above?

If we choose the test point (0,0), the coordinates zero and zero substituted in the inequality, this inequality will become:

$5(0) - 3(0) > 9$

Or, $0 > 9$ which is not correct.

From here we have that point (0,0) is not part of the solution, so the solution is the region beneath the line, not including the point on the line.

7) Without showing any work, sketch the graph of the following inequality:

$y + 2 < 0$

is equivalent with

$y < -2$

We take the region that is beneath the horizontal line $y = -2$ but not the points situated on the line.

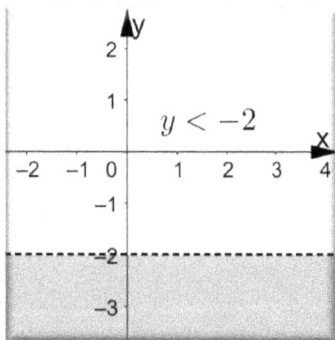

8) Show the solution region to the inequality: $3x + 4y > 8$

We transform the line $3x + 4y = 8$ into the slope intercept form $y = mx + b$

$3x + 4y = 8$

Step 1 we subtract $3x$ on both sides of the equation

$3x - 3x + 4y = -3x + 8$

$4y = -3x + 8$

Step 2 we divide by 4 on both sides of the equation

$\frac{4}{4}y = -\frac{3}{4}x + \frac{8}{4}$

$y = -\frac{3}{4}x + 2$

Here, the slope is $= -\frac{3}{4}$ and the y intercept is 2

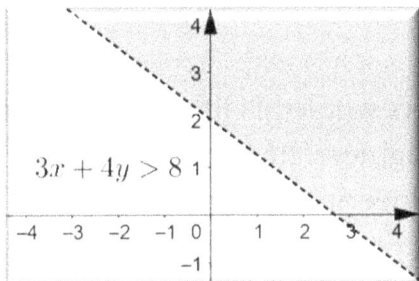

The test point is (0,0) in the original inequality.

$3(0) + 4(0) > 8$

$0 > 8$ not true,

So, point (0,0) is not part of the solution region.

$3x + 4y > 8$

9) The equation of the boundary is given. Determine the inequality which is represented by the solution region.

$3x - 5y + 15 = 0$

We use the test point (0,0) to see what condition we have to have so, the test point is part of the solution region. In this case the sign used has to satisfy the inequality

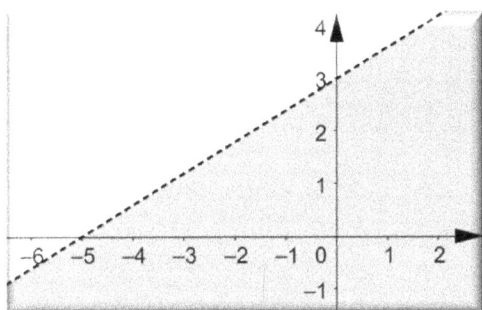

$3(0) - 5(0) + 15 = 0$

$+15 > 0$ it is true

So, the graph represents the inequality:

$3x - 5y + 15 > 0$

10) The point(s) which is NOT in the solution region of the inequality $5x - 2y > 5$ are:

a) (1,3)   b) (3,2)   c) (2,-3)   d(0,1)

a) $5(1) - 2(3) > 5$

$5 - 6 > 5$

$-1 > 5$ NOT TRUE   point (1,3) is not part of the solution region

b) $5(3) - 2(2) > 5$

$15 - 4 > 5$

$11 > 5$ true

c) $5(2) - 2(-3) > 5$

$10 + 6 > 5$

$16 > 5$ true

d) $5(0) - 2(1) > 5$

$0 - 2 > 5$

$-2 > 5$ NOT TRUE point (0,1) is not part of the solution region

ABOUT THE AUTHOR

Dr. Marcel Sincraian has been working with numbers whole his life as an Engineer, Accountant, Math and Physics teacher. While an Engineer, he got his Ph.D. in Civil Engineering. He participated in European engineering research projects in soil dynamics. He published papers in international Engineering Journals and international Conferences. He is a passionate, warm, and funny Mathematics and Physics teacher. He is always ready to share his knowledge in Math and Physics. Because of his teaching passion, he decided to help the students with a few math books, from Introduction to Calculus to pure Calculus. He is continually writing new Math and Physics books hoping to help students with other parts of Mathematics and Physics. His hobbies are, reading, traveling, camping.